NETLIFE:
Internet Citizens and their Communities

NETLIFE:
Internet Citizens and their Communities

CARLA G. SURRATT

Nova Science Publishers, Inc
Commack, NY

Editorial Production: Susan Boriotti
Office Manager: Annette Hellinger
Graphics: Frank Grucci and John T'Lustachowski
Information Editor: Tatiana Shohov
Book Production: Patrick Devan, Christine Mathosian, Tammy Sauter and Diane Sharp
Circulation: Maryanne Schmidt
Marketing/Sales: Cathy DeGregory

Library of Congress Cataloging-in-Publication Data available upon request

ISBN 1-56072-577-X

Copyright © 1998 by Nova Science Publishers, Inc.
6080 Jericho Turnpike, Suite 207
Commack, New York 11725
Tele. 516-499-3103   Fax 516-499-3146
E-Mail: Novascience@earthlink.net
Web Site: http://www.nexusworld.com/nova

All rights reserved. No part of this book may be reproduced, stored in a retrieval system or transmitted in any form or by any means: electronic, electrostatic, magnetic, tape, mechanical, photocopying, recording or otherwise without permission from the publishers.

The authors and publisher haven taken care in preparation of this book, but make no expressed or implied warranty of any kind and assume no responsibility for any errors or omissions. No liability is assumed for incidental or consequential damages in connection with or arising out of information contained in this book.

This publication is designed to provide accurate and authoritative information with regard to the subject matter covered herein. It is sold with the clear understanding that the publisher is not engaged in rendering legal or any other professional services. If legal or any other expert assistance is required, the services of a competent person should be sought. FROM A DECLARATION OF PARTICIPANTS JOINTLY ADOPTED BY A COMMITTEE OF THE AMERICAN BAR ASSOCIATION AND A COMMITTEE OF PUBLISHERS.

Printed in the United States of America

Contents

Preface	IX
List of Figures	VII
Introduction	1
A Beginner's Guide to Matrix Technology	13
Matrix Communities and Culture: An Overview	19
Understanding Community Structure: Social Institutions and Groups	45
Socialization and Personal Identity: The Creation of an On-Line Self	103
Deviant Behavior: Its Definition and Control	167
Social Inequality in On-Line Communities	227
An Interactionist View of On-Line Communities: Concluding Remarks	269
List of Abbreviations	277
List of Emoticons	279
Glossary	281
Bibliography	285
Index	297

LIST OF FIGURES

Figure	Page
1. The Matrix	x
2. Fidonet Connectivity	87

PREFACE

The central questions around which this book revolves are whether or not communication mediated through technology constitutes so-called real interaction and whether the establishment of real community is possible solely through electronic technology. Here real is defined as meaningful to the participants, as meaningful as face-to-face interaction, and the special case under consideration is communication through the computer Matrix.[1]

From a sociological perspective, these questions may be answered in the affirmative if computer-mediated interactions pass the general litmus test of 'community'. That is, computer-mediated communications (CMC), like face-to-face interactions, must simultaneously provide participants with: 1) the means to formulate and maintain individual, legitimated identities; and 2) social institutions, or problem-solving strategies, designed to meet basic community needs (i.e., establishment of a system of values and norms, socialization of new members, resolution of ambiguity, and personal accountability to the group). Failure to pass such a test means that, by definition, one or more of these needs remains unmet.

In order to answer these questions accurately, one must understand the specific cultural contexts of the interactions observed. However, the catch-all term for such contexts, the "Internet", is misleading because it encourages one to overlook specific on-line group contexts. The term was originally a shorthand for the concept of internetworking, an expression which was intended to be general and to indicate all processes which can and do happen through the use of computers, from information gathering to interpersonal communications to computers 'talking' to each other. Through popularization of the concept "Internet", people have erroneously subsumed a variety of technical systems and human intentions under a single frame. This is not difficult to do, given the degree of complexity of computer systems and the ease with which a term such as "Internet" grants simplicity to complex phenomena.

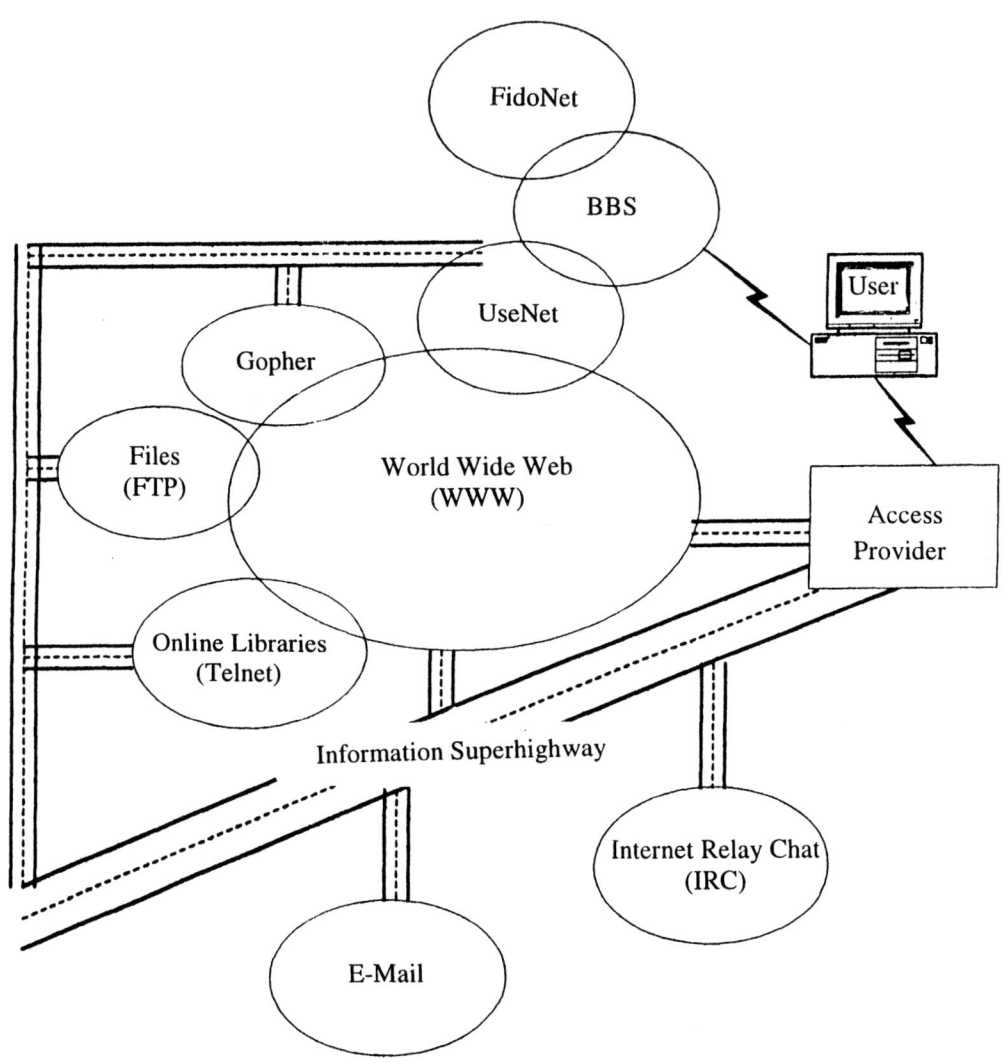

Figure 1. The Matrix

However, to continue to discuss the Internet as if it were a single, unified system of computers and cultural framework within which human communication can be understood only serves to exacerbate the problem. Such generalizations serve largely to cloud our understandings of both the processes of computer-mediated communications and the meanings of such communications because we continue to be denied an understanding of the specific cultural contexts in which they occur. Thus, many critiques of CMC are based largely on this popular definition of the Internet, and not on a clear understanding of the cultural frameworks in which such interactions are embedded.

At present, there is not one system, technical or cultural, which can accurately be called the Internet. A more accurate statement is that there are multiple systems and purposes within the rubric of internetworking, some of which entail the social properties described above and some of which, do not. At a basic level, the computer Matrix serves two distinct purposes and utilizes several formats for accomplishing those purposes. The first purpose is information gathering and/or providing, which can be accomplished through Gopher, FTP (file transfer protocol) or HTTP (hypertext transfer protocol), also known as WWW (World Wide Web, or simply The Web). The second purpose is human-to-human communication, (computer-mediated communication, CMC), which can be accomplished, in general, through Email (electronic mail), IRC (Internet Relay Chat), Usenet, Fidonet, and MUDs (multi-user dungeons, also referred to as MOOs--Muds, ObjectOriented and sometimes Tiny-or-TeenyMUD).[2]

Of the formats listed above, this book examines IRC, Usenet and Fidonet for three main reasons. First, these communication processes are multi-user and two-way. Second, they are publicly accessible through local gateway providers. Third, they are non-fantasy-based.[3] Bear in mind that while theses CMC formats share those basic properties, their techno-social histories are quite varied, and have resulted in distinct cultures. And each cultural framework requires that one employ a slightly different research methodology in order to most fully understand those contexts.

In general, the author employed the methodology of participant-observation, with the level of participation varying according to on-line community cultural norms. The IRC community allowed for, and even required, immediate, informal interaction (the reader will note the author's nickname, Marcia, sprinkled throughout sample interactions in the following chapters). Usenet and Fidonet on the other hand called for a more reserved data collection approach; one that tended toward observation rather than participation. In all, the author logged several thousand hours in these on-line communities and became "a regular" in several specific groups within those communities.

ENDNOTES

[1] John S. Quarterman's (1994) term for the Internet.

[2] These systems are simply representative of the most commonly used forms of CMC. That is, not all synchronous chat happens on IRC, not all information-based groups are Usenet newsgroups, and not all BBSs are on Fidonet.

[3] Multi-user, two way communications simultaneously allow for at least the possibility that a community might form (assuming a community requires more than two participants) and for an outsider to observe and discuss interactions. Communications provided through local gateways allow for the possibility of participants themselves setting and enforcing rules of behavior, rather than having a company decide policy. Finally, non-fantasy based communications allow participants to choose the extent to which they develop and can legitimate on-line identities similar to those found in face-to-face communications (MUDs and MOOs are most commonly used to play fantasy-based games in which, for example, contestants are warriors who attempt to slay dragons).

CHAPTER 1

INTRODUCTION

Recent debate about the meaning and impact of interaction through the computer Matrix, whether such interactions constitute real and meaningful behavior and can lead to the establishment of real communities, is the logical extension of the decades-long argument over the roles and impacts of other mass, electronically mediated communications. The basic arguments that critics of these forms of communication have made may be summarized as follows:

"All mass media in the end alienate people from personal experience and, though appearing to offset it, intensify their moral isolation from each other, from reality and from themselves. One may turn to mass media when lonely or bored. But mass media, once they become a habit, impair the capacity for meaningful experience." [van der Haag 1968, 5]

From this perspective, such communication is seen as a social problem, and ultimately not real communication at all. Although this argument has been made about print-mediated communications such as newspapers, books and magazines [Snow 1983], the critique in the recent past has focused on television. As noted by James Anderson and Timothy Meyer [1988], such critiques generally fall into the 'traditional' or 'effects' oriented theory of media and follow what Denis McQuail [1994] calls the social disintegration framework. That is, these critiques argue that people act "under the influence" of television, losing a sense of personal identity, real community and meaningful interaction with others.

For example, Bernard Rosenberg [1971] argued that, "television is an hallucinogen...the masses are victims of a merciless technological invasion that threatens to destroy their humanity...it anesthetizes people and they are tricked into believing everything they see on television." [pp. 4-6] Similarly,

Dwight MacDonald [1968] claimed that because of this effect, viewers become 'passive consumers' who lose their human identity because they are no longer related as members of a community. The results are two-fold; people lose their sense of self and must look to television for a confirmation of self-image [Mills 1968], and, ultimately, the television is a tool of political domination, of tyranny.

Perhaps the most forceful critique comes from Jacques Ellul [1991], who, in his analysis of the effects of mass, electronically mediated communications on personal identity and community, suggests that the world is not a global village for four reasons. First, people no longer know the personality, nature and ideology of the individual transmitting the information. Second, the human relationship is greatly altered when information is transmitted through technology because words are dissociated from the people who say them. Third, the information transmitted is not relevant to people's lives, in the sense that they cannot change the outcome. Finally, Ellul argues that "...only technicians use the mass media, and it is out of the question to penetrate their domain. The amateur has only his hobby. He is more eager to accept the information, because he feels he is taking part in the big game. The belief that anyone can send information is only a wish and a myth; not reality. . . . the proliferation of media seems to be fundamentally anti-democratic." [Ellul 1991, 353]

Most recently, communication mediated through computer technology has suffered criticism similar to that described above. (Note that such criticisms refer to the Internet as one, integrated system, rather than as a Matrix of varying techno-social networks.) In general, such critiques may be subsumed under one of four basic and interrelated arguments, which claim the following: first, interaction on the Internet is anonymous and thus unlike real social interaction; second, no one can be held accountable for the statements he or she makes; third, because there is no formal government, the Internet is in a state of anarchy; and fourth, because of these factors, the Internet does not constitute a real community.

Turning first to anecdotal critiques [Gravino 1995] made by people who participate in the Internet to varying degrees, one finds these arguments made regularly. Consider the following examples. Andre Bacard, a physicist at Stanford, comments on a tendency for excessive computer use in some people, "Their entire social world is vicarious, it encourages and cultivates psychosis in many people." [p. E2] James Donley, medical director of outpatient psychiatry at the University of Kansas Medical Center, comments, "It lets people have a fairly simple relationship, where you can kind of predict or dictate what happens. . .you don't have to put up with all the things about human relationships that are both exciting and unpleasant. It can be used as an avoidance of social relationships." [p. E2]

This notion of predictability and simplistic relationships is common among critics of so-called 'virtual' communities. Michael Heim [1995], a philosophy teacher and author, suggests that virtual community seems to be a 'cure-all for isolated people who won't give up their isolation'. He comments that "I know people in rural communities who hear wishful thinking in the phrase 'virtual community'. . . For many, real community means a difficult, never-resolved struggle. It's a sharing that cannot be virtual because its reality arises from the public places that people share. . . For many, the 'as-if-community' lacks the rough interdependence of life shared." [p. 3]

Other critics of the Internet suggest that there are a lot of 'fringe group' elements on the Internet, and that the lack of government combined with anonymity shield these users from law enforcement, unlike the real world. Walter Mossberg [1995], columnist for the Wall Street Journal, stated:

"Virtual communities are clubs for the technically adept. They must become more like open, democratic societies where people take responsibility for their actions and welcome the participation of others. The on-line community is not a digital democracy because discussion occurs in a climate where people hide behind assumed names and can smear others with impunity and censor those they don't like." [p. B1]

Supporters of this argument agree that are three main problem areas. First, there is anonymity; participants do not have to use 'real' names and thus it is impossible to figure out the 'true' identity of someone making claims. This lack of true identity supposedly makes it easier to engage in malicious activities, such as smearing people, concocting financial scams, and victimizing people sexually. The second problem is incivility; it is too easy to 'flame' people and on-line discussions too often break down into name-calling that would not be tolerated in a real-life meeting or social setting. There are no limits to expression except to be 'flamed' or canceled through vigilante justice.

Finally, there is the problem of censorship and intimidation; any participant may attempt to scare or gag people with whom he or she disagrees. In this regard, the Internet has been likened to talk radio; participants make subversive, anonymous arguments, the accuracy of which is hard to verify. People play on the fears of others and their willingness to believe everything that passes over electronic networks. Continuing this analogy between talk radio and Internet communication, Michael Halloran, a professor of communication, suggests that the computer medium itself may be the determining factor in the ability of participants to communicate effectively.

Halloran [1995] proposes that, in its current state of development, computer users are limited to words and emoticons.[1] Thus:

> "Internet flamers may simply be frustrated by their situation, and their frustration may express itself in the creation and excoriation of villains. In the virtual spaces of the airwaves and the Internet, I can become a transparent ego. The rantings of flamers on talk-radio and the Internet may be the spontaneous overflow of strong feelings, untempered by either the tranquility of recollection or the constraints of sociality. By placing people where they are morally alone and audible to thousands, these media may be releasing the dark feelings we once learned to conceal from all but our closest intimates, even from ourselves." [p. 11]

Two of these comments in particular bear a strong resemblance to the critiques made of mass communications through television. The first is the notion that interaction through computers is a social problem; it pulls participants away from so-called real social interaction and limits their ability to deal with others because they are accustomed to being able to manipulate situations involving the computer. The second is the idea that people are indiscriminate in their choices of what information to believe or what interactions to engage in simply because such information has passed through the computer medium. Traditional critiques of television claimed similar outcomes when they suggested that people are 'tricked into believing' everything they see on television.

Those who disagree with this basic assessment use the conceptual framework of symbolic interaction to suggest a different way to understand mediated communications. In general, symbolic interaction is the study of the making of meaningful behavior. Symbolic interactionists assert that meaning emerges from consensus among actors and is established in interaction. That is, meaning is not inherent, it is created and therefore unstable and problematic. [Brissett and Edgley 1975] Within this context, the self is defined as the meaning of the human organism; a meaning which is highly situational. [G. Stone 1970] The self is established by its activity and by the activity of others towards it. It is not inherent in the individual because the "individual" is a shared, interactive phenomenon; self is an outcome, not an antecedent of behavior. What one does establishes who one is, not vice versa; there is no useful distinction to be made between actor and action. Further, interactionists argue that human beings are primarily symbol users; all meanings are constructed through symbols, which are the primary cultural resource.

Another way of phrasing the statements above is to say that society is symbolic interaction. [Blumer 1969] That is, all human interaction is mediated by the use and interpretation of symbols. And, in order for such interaction to occur and be meaningful, each participant in that interaction must be able to take the role of the other through the use of the sign system of language. An important consequence of this use of language to bring meaning to the self and the world is that there is no ultimately 'real' world; the world and the self are 'real' insofar as they are known through language. [E. Becker 1962]

Given these basic premises, the symbolic interactionist response to critiques of mediated communications has generally hinged upon challenging the one-way, stimulus-response model of participants in those forms of communications. According to George Herbert Mead [1934], all meaning, including the meaning of the self, lies within the social act. The meaning of the self is a social process because it requires the individual to take the role of the other in interaction, thus making the self a social object. The meaning of the interaction is a social process because it involves both taking the role of the other and interpreting the other's acts in the context of a particular situation. Theorists of mass communications argue that this two-way process of interaction continues to be the case; the mediation of interaction through electronic technologies does not negate role-taking or the construction of meaning as a social process.

For example, Franklin Fearing [1964] adapts Mead's concepts of human interaction to mass communication and suggests that if communication is to occur between human beings, both parties must be implicated. That is, the meaning of the text communicated is not inherent in it, but must be constructed or interpreted locally by each participant in that interaction.

Regarding the notion of loss of personal identity, the symbolic interactionist approach has been to suggest that, while C. Wright Mills [1968] may be quite correct in suggesting that people look toward television for confirmation of identity, the manner in which this is done is what is important. That is, as suggested by David Manning White [1971], people may in fact "get partial answers from television to the basic philosophical questions-- who am I, why am I here, what is the meaning of my life." [p. 15] Robert P. Snow [1983] further suggests that, not only do people use mass media as an information source for development of new identities, they also use it as a source of validation for existing identities. The point of distinction between the traditional approach and the symbolic interactionist approach is how one views the participants. In the traditional approach, participants are 'tricked into believing' everything on television; in the interactionist approach, people are highly selective in terms of the formats and contents they choose to pay attention to.

Concerning the critique of mass media as contributing to a loss of community, Lewis Anthony Dexter [1964] suggests two ideas. First, he argues that mass communications have allowed people to achieve consensus on issues with 'strangers' as well as 'neighbors'. His second and interrelated argument is that, because electronically mediated communications break down traditional barriers of time and place, the distinctions between primary and secondary communications have become blurred; "in specific, actual cases, the difference between secondary and face-to-face communications is hard to work out." [pp. 9-10] Similarly, McQuail [1994] argues that what are generally thought of as mass media can also be used as social activities:

"There is little doubt that much media use is sociable, contrary to the image in mass society theory of the audience as made up of isolated individuals, in an atomized society. While it is true that the rise of mass media increased the possibility for solitary attention to more channels of public communication (radio, newspapers, cinema, etc.) and reduced the dependency of the individual on other people for human contact, media use is in practice as sociable or as solitary as a person wants it to be." [p. 308]

The interactionist framework has some important implications for our understanding of communications mediated through the computer Matrix. According to Allucquere Roseanne Stone [1991], the virtual on-line communities are 'passage points' for collections of common beliefs and practices that unite people who are physically separated. Such communities have evolved in the fourth epoch of communications, each epoch being marked by communications mediated through increasingly complex technologies, as follows: first, the text epoch, originating in the 1600s; second, the electronic communication and entertainment media epoch, beginning in the 1900s; third, the information technology epoch, beginning in the 1960s; and fourth, the virtual reality and cyberspace epoch, beginning around 1984. Stone argues that each of these epochs is marked by the progressively hidden body, first by conservative clothing, and then by increasing spatial privacy. Thus, as communications become increasingly asynchronous, the notion of 'where' and 'when' something occurs becomes problematic; *the body becomes more physical while the "I" becomes more textual* (italics mine).

Given an "I" which is increasingly textual in nature, Snow [1983] examines the role that electronically mediated communication plays in the symbolic construction of the self and other meanings through which people live their lives. In short, he concludes that mediated communication and interpersonal communications, if the distinction may even be made, follow the same social-psychological procedures. That is, language is used in order to take the role of the other and assess one's own self-conception based upon that other. Whether that 'other' is physically present is largely irrelevant.

Despite the intensity of concern over and debate on the subject, much of our collective knowledge about the extent to which CMC functions as a forum for meaningful interaction and serves as a human community stems primarily from anecdotal information gleaned from participants. Scholarly research on the subject remains relatively limited. Often, these first-hand accounts of experiences on-line make the case for the meaningfulness of such interaction implicitly by assuming that it is self-evident that communication mediated through computers is just like face-to-face encounters.

For example, regarding gender-specific spaces in CMC, users say "It's an attempt to connect, not to show off" and "I see us struggling with being male in this society, and dealing with pain and emotional hurt". Some participants are extremely committed to the reality of 'virtual relationships' as the following quotes suggest: "On Women's Wire, people speak my language. It's not that men are not present, it's just that they're not central. It makes a world of difference"; "I use it to feed my soul.", and; "I'm fascinated with what women deal with and hooked on a level of emotional involvement and information I've never been able to find anywhere else in my life." [Ness 1994, E1]

In other instances, the comparison between face-to-face encounters and Matrix encounters is more explicit. Consider the following examples: Yvete

Colon, a social worker who is launching her second virtual group for online therapy says, "It actually works better that they never see each other. Since they're not busy watching each other, they concentrate more on themselves. The level of self-disclosure and trust is far greater than traditional groups." [Copilevitz 1995, D5]. Lorraine Harrington, middle school student in Colorado involved in a 'kid-space' portion of the Matrix comments, "It's a lot easier to communicate on-line. We talk about life. We talk about anything and everything". Another student says "It's easier to talk to girls on the Internet than in school. Sometimes I can't talk well in person. It doesn't come out like I want." Another says, "People treat you more like an adult than they do in person." Despite comments such as these, educators at the school wonder whether CMC is building social skills or providing an escape from difficult personal interaction. [McCartney and Rigdon 1994, B1]

David Woolley and Michael Hauben, both contributors to Computer-Mediated Communications Magazine, comment on their experiences with a conferencing system named Plato and a Usenet group named nyc.general, respectively. Woolley says that while the Plato system was ". . .designed for computer-based education, its most enduring legacy is the on-line community spawned by its communication features. The Talkomatic portion of the system was an instant hit. There was no way to contact a specific person to let them know you wanted to talk, so it was more like a virtual water cooler than a telephone substitute. People would hang out in a channel and chat or flirt with whoever dropped by." [Woolley 1994, 7] And Hauben [1995] discusses nyc.general, a Usenet newsgroup that "demonstrates a friendliness of a good neighborhood in the midst of an ever growing city. . ." and likens the group to the "Greenwich Village 1963" model of community proposed by Sally Banes in a community study in which she suggested that community is not a static social form that is disappearing, but that new, dynamic and overlapping forms of small-scale networks have arisen. [p. 12]

Elderly people involved with SeniorNet [McCartney and Rigdon 1994] comment, "It's like a group of people sitting around a big table having coffee and kidding each other."; "It is a way for homebound people to get out and see friends."; "Even friendships that remain limited to cyberspace seem as good as the real thing."; and another says, "It doesn't matter. Your disabilities, the color of your skin. It makes no difference to these wonderful people." [p. B1] Two people who met through the Matrix and then got married explain, "We are having some difficulty in getting people to understand that we know all about one another." He popped the question on-line and she immediately responded 'yes'. [Love On the Line 1995, A4] Of course, this sense of real community applies as well to what are commonly referred to as 'fringe groups'; these groups are organized around issues ranging from gay rights to environmentalism to Ku Klux Klan-type efforts and they, too, report that they interact on-line to seek camaraderie and validation of their views.

The most extensive anecdotal discussion about meaningful interaction and the Matrix as a human community is The Virtual Community: Homesteading on the Electronic Frontier [Rheingold, 1993]. In this work, Rheingold describes in detail his experiences as a member of the Whole Earth 'Lectronic Link (WELL) on-line community. While he acknowledges that critics of computer-mediated communication may have some valid concerns regarding the extent to which such interaction is a substitute for face-to-face interaction, he makes the case that there is something very real about on-line community, as the following quote suggests: "There's always another mind there. It's like having the corner bar, complete with old buddies and delightful newcomers. . . and fresh graffiti and letters, except instead of putting on my coat, shutting down the computer and walking down to the corner, I just invoke my telecom program and there they are. It's a place." [p. 24] It is interesting to note that Benedict Anderson's Imagined Communities [1983] makes a similar case; all communities are imagined and must be commonly imagined in the minds of people if they are to exist at all. With these arguments in mind, one could suggest that Matrix community may be as real as any other.

While the scholarly work on the subject is still limited, a few studies have been done of CMC participation which make the case for the existence of Matrix communities. Voices from the WELL: The Logic of the Virtual Commons is one such example. [Smith 1992] The author begins with a basic symbolic interactionist discussion of the creation of the self and argues that participants in the WELL, though their means are limited, still create personality and identity for themselves using text alone. The fact that WELL interaction is aspatial, asynchronous, acorporal and thus relatively anonymous does not mean that identity is not present. Identity remains because people invest expectations and evaluations in other's user i.d.'s. If people wish to maintain their social network and knowledge capitals and the status that derives from those, they will continue to act in-line with their established identity or suffer the consequences. He concludes that, "despite the unique qualities of virtual worlds, people do not enter new terrain empty-handed. . .

Interaction in virtual space is not fixed, determined or easily controlled or directed. If we are to understand this new terrain of interaction, detailed work should be done that addresses how virtual communities form and mature, and how those relations are similar to and different from social relations in 'real-life'. " [Smith 1992]

In The Sociology of Culture in Computer-Mediated Communication: An Initial Exploration, Elizabeth Lane Lawley [1995] makes a similar argument for the community nature of on-line communications. Utilizing Bourdieu's [1984] concepts of culture and symbolic capital, Lawley argues that computer-mediated communications (CMC) does constitute a field worthy of scholarly analysis. She suggests that participants develop differing statuses relative to one another, and those positions endow CMC participants with a history. Further, she states that the linguistic conventions unique to CMC, combined with systems of netiquette (net etiquette) combine to form collectively effective methods for the regulation of behavior within the group.

In The Network Nation, Starr Roxanne Hiltz and Murray Turoff [1993] offer yet another analysis of CMC from the symbolic interactionist perspective. According to the authors, because communication channels are narrowed, participants lose all visual and some audio cues regarding social category or 'type' of other participants, facial expression, eye contact, body language and so forth. Thus beginners in CMC experience the equivalent of culture shock because the known rules for combining data from channels do not apply. For example, they cannot acknowledge and yield turn-taking in conversation without difficulty. However, Hiltz and Turoff do note that new participants begin to learn the new rules of interaction almost immediately and within a short time (30 minutes) report greater facility of use. The authors argue that this narrowing of communication channels does not mean that meaningful interaction is not taking place. In much the same way Smith argues that WELL participants learn to construct identities, Hiltz and Turoff's participants learn to construct new means of communication.

Very recently, a body of scholarly work has begun to emerge which challenges even the assumed distinction between 'virtual' and 'real' community. Nancy Baym, who has researched a Usenet newsgroup dedicated to the discussion of soap operas points out:

"Too much of the work on CMC assumes that the computer itself is the sole influence on communicative outcomes. This is sometimes called the 'cues filtered out' approach because, in this perspective, the computer is assumed to have low social presence and, therefore, to deprive interactants of salient social cues. . . .The presumed lack of contextual cues and feedback is seen as producing several interrelated communicative outcomes. Interactants gain greater anonymity. . . participation is said to become more evenly distributed across group members. . . .The anonymity and lack of socioemotional information is taken to erase norms for interaction. . .people become less socially inhibited and more likely to be rude. . . "[pp. 139-140]

She argues that such critiques are inaccurate, and asserts that the social realities of identity, relationships and community are created through interaction as CMC participants draw on language and the resources available to make messages that serve their purposes, just as they do in face-to-face communications.

This work is broader in scope than any of those outlined above. It discusses all of the essential sociological elements of three very large on-line communities in which millions of people around the world participate everyday. The reader is invited, in the chapters that follow, to gain a deeper understanding of life in these on-line communities. Why do individuals become involved in on-line interactions? How do individuals come to feel as if they belong to a group, a community? What forms do such communities take, and how do their members maintain a sense of order and group cohesion? Tour the Matrix communities of IRC, Usenet and Fidonet, and make sense of this phenomenon from a social-psychological perspective.

ENDNOTES

[1] Emoticons are symbols generated through combining keyboard symbols into forms that resemble human actions or expressions. The most common emoticon is the 'smiley', which can take on many shapes, but is usually indicated as :-)

CHAPTER 2

A BEGINNER'S GUIDE TO MATRIX TECHNOLOGY

In order to appreciate the distinct cultures of IRC, Usenet and Fidonet, one must have a basic understanding of the techno-social origins and evolution of the Matrix. What follows is an abbreviated history of the evolution of these three computer networks, beginning with a discussion of the development of the Internet 'backbone', ARPANET, as it relates to the Internet Relay Chat (IRC) network, continuing with a discussion of a closely related network, Usenet, and concluding with a discussion of the last network, Fidonet.

According to Howard Rheingold [1993], "The essential elements of what became the Net were created by people who believed in, wanted, and therefore invented ways of using computers to amplify human thinking and communication. And many of them wanted to provide it to as many people as possible, at the lowest possible cost. Driven by the excitement of creating their own special subculture below the crust of the mass media mainstream, they worked with what was at hand. Again and again, the most important parts of the Net piggybacked on technologies that were created for very different purposes." [pp. 66-67]

In 1969, what was to eventually become the Internet began as a command-and-control network conceived by The Rand Corporation and sponsored by the U.S. Department of Defense. Named ARPANET, it was originally designed as a deliberately decentralized system that would allow U.S. authorities to successfully communicate after a nuclear war. In its infancy, this network consisted of only four computers. In 1971 there were 15 computers on the network; by 1972, there were 37. However, despite the fact that the network was designed for long-distance computing for research and defense purposes, by the second year of operation, ARPANET's users had transformed the network into a social phenomenon. The main traffic on ARPANET was not long-distance computing, but personal e-mail messages and bulletin board news services.

ARPANET continued to grow throughout the 1970s and 1980s, as many different social groups found themselves in possession of powerful computers. It was fairly easy to link these computers to the growing network-of-networks because the software was public-domain, and the basic technology was decentralized by its very nature. This network-of-networks eventually became known as the Internet.

Today there are tens of thousands of nodes (linkage points) in the Internet, which spans more than seventy-five countries and has a population estimated between four and ten million people. The Internet's pace of growth in the 1990s is estimated to be between ten and twenty percent per month, with the number of 'host' computers doubling every year since 1988 to over nine million in 1996. The total number of connected networks increased from about two hundred in the early 1980s to over seventy-five hundred a decade later, as the Internet continued to move out of its original base in military and research institutions and into elementary and high schools, public libraries and the commercial sector. [Sterling 1993]

Internet Relay Chat is a teleconferencing system which uses standard Internet protocol to allow communications among users in real time across the backbone described above. The network is made up of clients, which are end users, and servers, which are the central points that connect clients together and allow for connections among other servers. Each client is known to all servers by a nickname, the username and the server to which the client is connected. [Oikarinen and Reed 1993] In order to maintain the smooth functioning of the IRC network, a special class of clients exists, called operators, which perform maintenance functions on the network. These operators are granted powers that are both dangerous and necessary to the functioning of IRC. Finally, the spaces in which this real time communication takes place are commonly called channels. A channel is a named group of one or more clients which all receive messages addressed to that channel. A channel is created implicitly when the first client joins it, and the channel ceases to exist when the last client leaves it (disregarding, for the moment, robots which can maintain channels). Channel operators are those who are considered to 'own' channels, and these operators are granted certain technical powers that allow them to maintain control of their channels.

IRC protocol was first implemented as a means for users of Bulletin Board Systems to chat in real time. Jarkko Oikarinen designed the original client-server system in 1988, and it was first tested on a single machine with fewer than 20 clients. IRC gained international fame during the Persian Gulf War, when news updates from around the world were funneled into a single channel which clients could connect to in order to receive and discuss the news. Since that time, IRC has also been used as a low cost means for spreading and discussing eye witness accounts of the Russian coup and California earthquakes in real time. [IRC for the Newcomer FAQ 1995]

The original IRC network was established in Finland, but IRC has now spread worldwide. However, due to political disputes, there are now multiple parallel 'threads' within IRC, the two largest of which are EFNet (Eris Free Network) and Undernet. On average, there are a little over 9000 clients and around 100 servers on EFNet, while Undernet averages around 1400 clients on about 30 servers.

Usenet is not synonymous with the Internet as described above, although information in Usenet newsgroups may pass through an Internet network. Further, Usenet is not generally thought of as a network in the same sense that the Internet is a network, though there is some disagreement on this point among Matrix experts. Usenet was originally established in 1979 by Tom Truscott and Steve Bellovin in order to transfer information between Duke University and the University of North Carolina. [alt.culture.usenet FAQ 1995] During its early years, Usenet news was transmitted by UUCP, unix-to-unix connection protocol, and NNTP, a protocol that would allow the transmission of Usenet news over ARPANET, had yet to become widely used. This distinction between protocols is important, because it explains how the Internet and Usenet came to be somewhat interconnected, and why they are not synonymous.

Information which travels through one of the networks that makes up the Internet travels by IP, or Internet Protocol, and often times by TCP/IP, or Transmission Control Protocol, which is one of the protocols associated with IP. Because these protocols (UUCP and TCP/IP) differ from one another, computer users have access to different services depending upon the protocol. More importantly, these technological differences, which were inspired by a desire to transmit different types of information and services in different ways, lead ultimately to the evolution of different cultures within computer-mediated communications.

Unlike the Internet, Usenet has no 'backbone' in the strict sense. That is, Usenet has the form of a graph. On this graph, one can choose to emphasize a set of lines forming a path through the computer hosts. So, what early participants in Usenet referred to as a 'backbone' was simply a group of hosts whose administrators agreed to form such a connected set and that this set would carry all Usenet traffic as promptly as possible. [The Great Usenet Renaming FAQ 1994] Originally, the dominant means of transmitting newsgroup news was UUCP through long-distance dial-up lines. Thus, the 'backbone' consisted of those sites that were well connected. Today, however, even the smallest Usenet site has connectivity surpassing that which the former 'backbone' sites used to enjoy. Unlike the Internet, which flourished largely because of government funding, Usenet was started in a university setting. And while many of the Usenet hosts today are universities, most sites are commercial entities. Thus, the origins of Usenet are clearly more haphazard than those of the Internet, and the cultural framework that developed is a reflection of such ambiguity. [What is Usenet FAQ, 1995]

So, Usenet is neither synonymous with the Internet, nor is it a network in the strict sense. Sterling [1993] describes it as "an enormous billowing crowd of gossipy, news-hungry people, wandering in and through the Internet on their way to various private backyard barbecues. Usenet is not so much a physical network as a set of social conventions." Usenet is a world-wide, asynchronous, distributed discussion system. It consists of a set of newsgroups with names that are classified hierarchically by subject. Articles, or messages, are posted to these newsgroups by people on computers with the appropriate software. These articles are then broadcast to other interconnected computer systems via a wide variety of networks. Usenet is available on a wide variety of computer systems and networks, but the bulk of its traffic is transported over the Internet or UUCP. [What is Usenet FAQ 1995] Today there are about 2,500 separate newsgroups on Usenet, and their discussions generate about 7 million words of typed commentary every single day. [Sterling 1993] This is quite a departure from the original design estimates of Usenet, which estimated a maximum traffic volume of two articles per day. [Bellovin 1995]

In contrast to the Internet and Usenet, Fidonet is a point-to-point and store-and-forward email wide area network (WAN) which uses modems on the direct-dial telephone network. It uses a unique protocol, separate from those used by the Internet and Usenet, although there are gateways between Fidonet and the Internet, and Fidonet is capable of carrying Usenet newsgroups. [Bush 1993] While the funding distinctions between the Internet and Usenet are somewhat important to understand in terms of their cultural development, funding is the key issue which distinguishes Fidonet culture from both the Internet and Usenet.

Fidonet was begun almost accidentally in 1984 by Tom Jennings, who wished to move messages from his bulletin board service (BBS) in San Francisco to the BBS of a friend, John Madil, in Baltimore. Jennings wrote a computer program which was designed to move these electronic messages, during a designated hour in the middle of the night when phone rates are lowest, from his node to the one in Baltimore and back.

According to Jennings [1985], the original purpose of Fidonet was "to see if it could be done, merely for the fun of it, like ham radio. But it quickly became useful; instead of trying to call each others' boards up to leave messages, or expensive voice phone calls, Fidonet messages became more or less routine." [p. 1] As more nodes were added to the system, the rudimentary computer programs were ineffective at handling the increasing level of traffic. "With 30 systems, coordination became difficult: instead of a simple voice phone call to the very few system operators (sysops) to straighten out problems like modems not answering, wrong numbers, clock problems, etc., it took days to get the slightest problem repaired." [Jennings 1985, 1]

Unlike the Internet, which was originally government sponsored and is now largely in the hands of commercial enterprises, and Usenet which is sponsored by universities, commercial activities, and some private individuals, Fidonet (especially in the US) is almost entirely financed by private individuals. Because of this, minimization of modem and telephone time has been the principle driving force behind the design of the data transfer protocols; protocols are designed for maximum efficiency. [Bush 1993]. In addition, the emphasis on cost-minimization lead to a very decentralized organization of Fidonet. A list of all nodes in the network is updated and distributed weekly to every node; point-to-point data transfers are always possible. However, because Fidonet's email store-and-forward capabilities are specified in the network's minimum standards for protocol, email tends to be routed through a world-wide hierarchy on a somewhat ad hoc basis. That is, Fidonet email moves up through a hierarchy from a point (a sub-node not technically within the Fidonet system) to a node (a computer listed on the public nodelist which exchanges mail with other nodes) to a network (all of the nodes for a particular geographical area) to a region (a collection of networks) to a zone (a set of regions, usually corresponding to continents) and back down again.

Fidonet provides the most publicly accessible and lowest-cost email and enews service; in April 1993, the nodelist stood at 20,000. Of those 20,000 nodes, about 59% are in North America, 30% are in Europe, 4% are in Australia and New Zealand, and 7% are in Asia, Latin America and Africa. What this means in terms of the development of Fidonet culture is that, unlike the Internet and Usenet, Fidonet "has primarily been owned and operated by end-users and hobbyists more than by computer professionals." [Bush 1993] The variety of users combined with the self-sufficiency of each node in the network has combined to create a political climate in Fidonet that is substantially different from that of the Internet and Usenet.

CHAPTER 3

MATRIX COMMUNITIES AND CULTURE: AN OVERVIEW

Defining 'community' has always been a difficult task. Even for individuals who specialize in the study of community, a clear, meaningful and lasting definition has proved elusive. The term is one which is complex and abstract; research on the subject mainly consists of case studies of particular settings which researchers assume to be communities because the basic question, 'what is a community?' is a question with many possible answers.

One possible definition of community is a people who interact within a limited territory and who share a culture. Or it might be defined as the largest social organization whose patterns make a significant difference to the individual's actions; the social organization within which all other social organizations exist. But exactly what constitutes territory and interaction, and how one can determine which social organization is the largest one which makes a 'significant' difference to the individual? With the advent of the telephone, the television, the computer, the multinational corporation and so forth, is any system of social relations bound by a geographic limit other than a global one? According to Colin Bell and Howard Newby "...out of community studies, there has never developed a theory of community, nor even a satisfactory definition of what a community is. The concept of community has been the concern of sociologists for more than two hundred years, but even a satisfactory definition of it in sociological terms appears as remote as ever." [1974, xiii]

This dilemma regarding which social groupings may legitimately be called communities has its roots in the work of Ferdinand Tonnies. Tonnies defined community (gemeinschaft) as a type of social organization in which people are bound closely together by kinship and tradition; any social setting in which people form what amounts to a single primary group. He contrasted gemeinschaft with the idea of 'gesellschaft', meaning 'association'. Associations are types of social organizations in which people typically have weak social ties and a great deal of self-interest. People are motivated by their own needs and desires rather than a desire to advance the well-being of everyone. In this context, modernization is equated with the progressive loss of human community, which provided personal ties, a sense of group membership and loyalty within small communities. The social world of the community was circumscribed in space as well as in its way of life; modernity erodes such possibilities.

Tonnies' definition imbued the study of community with a focus on physical place, moral undertones and a romanticism for the past, by suggesting that gesellschaft is a kind of social life which is cold and impersonal. By contrast, gemeinschaft is warmer and more affectionate. This idea is reflected in the current debate about what a community is; the role of physical space or territory in the development of group cohesion is of paramount importance. While many would agree with Tonnies' focus on bounded locale and the sense of solidarity often presumed to result from such circumstances, others argue that the role of physical territory, as traditionally viewed, is increasingly irrelevant.

Within the field of community sociology, there are definitions of community which attempt to alter the definition of space, and thereby allow for new forms of community; forms that more accurately reflect how people live today. For example, Gerald Suttles [1972] advances the idea of 'community of limited liability'. He argued that, in order to understand community forms today, one must focus on the intentional, voluntary and partial involvement of the members. Communities are no longer defined solely as locality-bound individuals who 'must' work together. Rather, participation in community today consists of voluntary associations of individuals who focus on specialized and often limited issues. Such communities have 'official' identities, meaning they are recognized by both members and non-members as 'real' groups; communities of limited liability have external advocates and/or adversaries. It is this voluntary participation, combined with recognition from 'outsiders', rather than physical space limitations, that constitute community boundaries today.

Similarly, David Minar and Scott Greer [1969] propose that while community used to refer to a physical concentration of individuals in one place, one must now look at community as the social organization among a concentration of individuals. They borrow from George Homans' [1950] concept of the 'human group' and state that the root of all community is organization. Community begins with the interaction and mutual modification of behavior. Such action eventually becomes patterned, and through such mutual modifications emerge shared perspectives and commitments to the 'place' and its group. These shared perspectives represent the group's culture; the agreed upon definitions of what the world is and should be like. They ultimately conclude that what binds a community together is a state of mind on the part of its members. Community is not a place, it is the set of social identifications that emerge from interaction. And out of such interaction, both cooperation and conflict can and do result. Thus, it is inaccurate to suggest that older forms of communities (gemeinschaft) are necessarily circumscribed social spaces which result in cooperation, while modern day associations (gesellschaft) are large, impersonal social spaces that necessarily result in conflict.

Benedict Anderson [1991] takes this notion of community as a state of mind one step further. He suggests that all communities are imagined communities, and that they must be imagined, because ". . .the members of even the smallest. . .will never know most of their fellow-members, meet them, or even hear of them, yet in the minds of each lives the image of their communion. . . .Thus, communities are to be distinguished, not by their falsity/genuineness, but by the style in which they are imagined." [p. 6]

These later conceptions of community, which attempt to redefine space in order to take into consideration the rise of a global order, reflect the symbolic interactionist perspective on community. In essence, this perspective suggests that:

". . .the world in which the self exists is no less a fiction than is the self: both are created from symbolic stuff. . . for the human animal, space is cultural--it has to be conceived within the cerebral cortex, and not be merely 'out there'. The boundaries of the space that we live in do not have to be in our immediate experience; in fact, they may never be. But we must be able to imagine them, so that our world has compass points. . . .The self, in order to exist, has to 'be' in relation to something else. The individual has to exist in reference to past, present and future places." [Becker, 1962, pp. 53-54]

This symbolic interactionist view of 'space' goes to the very heart of what community means in computer-mediated communications. As Steven Jones aptly summarizes,

> "In assessing the history of community studies, one finds that space was understood less as socially produced and more as that which produced social relations. So, for instance...threads running through definitions of community in the sociological study of community include territory, social system and sense of belonging, social interaction based on geographic area, self-sufficiency, common life, consciousness of a kind, and possession of common ends, norms and means. . . On the one hand, it (CMC) appears to foster community, or at least the sense of community, among its users. On the other hand, it embodies the 'impersonal' communication of the computer and of the written word...Traditional life, supposedly, was marked by face-to-face intimate relationships among friends, while modern life is characterized by distant impersonal contact among strangers. Communities are defined as shared, close and intimate, while societies are defined as separate, distanced and anonymous."[1995, pp. 18-23]

In the symbolic interactionist tradition, Jones argues that CMC is socially produced space. Communities formed by CMC can be defined as 'incontrovertibly social spaces in which people still meet face-to-face but under new definitions of both 'meet' and 'face'', In that sense, cyberspace hasn't a 'where'. Rather, the space of cyberspace is predicated on knowledge and information, on the common beliefs and practices of a society abstracted from physical space; the community is social interaction or social networks. Ultimately, CMC can be understood to build communities because it renders the idea of physical place irrelevant.

What does all of this mean for the on-line communities examined here? It means that, though the circumstances are different from communities built through face-to-face interaction, members of these on-line communities are still capable of developing and sustaining what historically have been the basic elements of culture; a generalized system of values, beliefs, norms and symbols.[1]

THE INTERNET RELAY CHAT COMMUNITY

As explained in Chapter 2, EFNet was the first IRC network and was developed in 1988. By 1990, however, political differences among key people in the IRC hierarchy resulted in the formation of a number of other real time chat networks which have since, for the most part, withered away. It was not until late 1992 that the Undernet began to take shape:

"In October, 1992, people had already started feeling the effects of user overload on EFNet, not to mention many other factors. In an effort to strike out new territory in cyberspace, many 'groups' of individuals had relentlessly attempted to sustain the onslaught of the forbidding isolation, and start up a server of their own. Many failed in their ventures, many just got bored with it and many just got shut down forcibly by paranoid system administrators (sysadmins). Through the scattered outposts emerged two groups. . .one in the U.S. and the other in Europe-Canada. . . . In the last week of December, 1992, the two nets merged. Ever since, a wide variety of servers have appeared and disappeared from the face of the Undernet, but the ones that have held on have always carried forth the spirit. In the face of all odds, against an almost non-existent userbase, they have clung to one another, each operator giving encouragement to the other, trying to lighten up those gloomy days when there were hardly any users. . .the Undernet lives on. . . "[Mirashi, p. 3]

The significance of statements such as those above, which represent the official history of the Undernet, cannot be overemphasized. The individuals who have worked to establish the Undernet pride themselves on their historic split from what they consider to be a largely 'overrated' EFNet system, and much of the cultural identity of Undernet springs from this split. That is, Undernet, as a culture, defines itself in relation to EFNet and vivid descriptions such as the one above serve to reflect this pride in identity.

What little written history is available regarding Undernet tends to revolve around exploding the myths that 'EFNetters' have supposedly created about the Undernet in order to discredit its worthiness as a system. For example, according to the Undernet IRC FAQ, the top six derogatory myths that EFNetters attempt to perpetuate regarding the Undernet are: 1) the Undernet is lame; 2) the Undernet has no users, or, the Undernet has more servers than users; 3) the Undernet was formed by people who couldn't get to be IRC operators (ops) on EFNet; 4) the Undernet servers are run on-the-sly, with no approval from system administrators; 5) Undernet solutions to (computer) protocol work only on a small net--they do not work on large nets; and 6) the Undernet operators are clueless and know nothing about routing.

Against such accusations, 'Undernetters' attempt to redefine themselves and their culture in a positive light:

> "The Undernet consists of some very highly motivated and dedicated people, enthusiastic to make a success of their venture. EFNet is becoming more and more politicized. . .With the various EFNet administrators forming groups against one another, the amount of cooperation between them has become almost nil. The time which they could spend in serving you, the user, is instead spent in endless bickering. Undernet is a net where the operators are friendly, easy going folks, and are always happy to help the user. Abusing users is highly frowned upon, and Undernetters follow a certain Undernetiquette. The goal of the Undernet is to provide a better environment for its users, with protection against malicious users who try to work against IRC principles." [Undernet IRC FAQ]

Thus, members of the Undernet have created, from a sociological perspective, a 'we' and a 'they' which they then utilize in order to legitimate their own social institutions and identities. The Undernetter argument is that their community way-of-life revolves around two basic values. First, they pride themselves on the friendly atmosphere they have worked to create. And second, they are dedicated to a more 'fair and efficient' communications system. To those ends, members have developed: 1) rules of netiquette (net etiquette); 2) an IRC-specific lingo; 3) a system of emoticons and action commands to convey body language; and 4) computer protocol enhancements that lessen the likelihood of both netsplit and lag--two communications problems that have the potential to severely disrupt the ability of participants to maintain personal identity and group cohesion.

The basic IRC rules of netiquette, or norms of behavior, are as follows: 1) do not 'flood' the channel with text; 2) do not use beeps in messages; 3) do not use profanity in public messages; 4) do not harass other users with unwanted messages or comments; and 5) do not engage in 'destructive' behavior which reduces the functioning of IRC (running clonebots, floodbots and nick colliders).[2] In general, these rules serve to create a 'level playing field' socially. Rules two, three and four represent the philosophy that all users have the right to participate in the conversation free from harassment and offensive behavior; they serve to maintain the light-hearted and easy going atmosphere.

Rules one and five are intended to enhance social equality by simulating technical equality. It is understood by IRC system administrators that users represent a wide range of technical abilities, and that the technical limitations of users can result from both individual knowledge of computers and physical limitations of the systems being used. So, for example, there is a rule against 'flooding' a channel with text because this can easily cause users with slow modems to experience lag and/or netsplit. In addition, there are rules against running clonebots and floodbots for the same reasons; not only can slow users be forced off the system, but unsuspecting 'newbies' (new users) can be lured into running such bot programs without fully understanding the technical implications. When this occurs, the likelihood of channel disruption is very high. Finally, programs known as 'nick colliders' are prohibited because they, in essence, take over an individual's identity by assuming his or her nickname (nick). Because nicks cannot be registered and because they are the first form of identification to other users, 'jacking' someone else's nick, especially through use of a nick collider, is a serious offense.

Members of IRC have also developed a lingo which is simultaneously designed for maintaining both communication efficiency and the identity of the individual as a member of a given channel. For example, 're' is a 'repeat hi' and is used when one leaves a channel momentarily and then rejoins. The use of 're' saves time and space because users are not required to go through the elaborate welcomes that usually occur when a 'regular' has joined a channel for the first time that day. In addition, the use of 're' is an acknowledgement on the part of others that they are aware that the user was in the channel only moments ago; it is easy, when one lacks a visual presence to others, to be forgotten in the conversation relatively quickly after one has left the room. 'Re' acknowledges that the group recalled the user's presence even during his or her absence, and such acknowledgement is a very reassuring thing indeed.

Similarly, 'brb', 'bbl' and 'bbiaf', which stand for 'be right back', 'be back later' and 'be back in a few', respectively, are used when an individual plans to leave the channel for only a short time. Users, particularly channel regulars, utilize these statements to both announce their intentions to leave the group and to emphasize that they will not be gone long. Such statements are commonly used to maintain group or conversation cohesion in a technically uncertain atmosphere. That is, IRC users are aware that technical problems are a constant threat, and that, at any time, any individual may have to exit the channel to correct such a system problem. Brb, bbl and bbiaf are efficient means of communicating to others in the group that one must leave, often immediately, through no fault of one's own, and that one intends to rejoin the conversation as soon as possible. Basically, this allows others involved in the conversation to maintain a sense of identity and cohesion; their selves have not been simply dismissed by the other.

There are many additional examples of IRC lingo, such as 'lol' and 'rotfl' which stand for laugh out loud and rolling on the floor laughing. Such statements are very commonly used, and reflect the 'just fun' atmosphere of IRC. Other examples of efficiency-oriented lingo are: 'u' for you; 'y' for why; '2' for to or too; 'b' for be; 'r' for are; and 'c' for see. Sometimes, these shorthands are combined, for example in 'oic' which stands for the exclamation, 'oh, I see!'.

IRC participants have also developed and are constantly experimenting with emoticons, a system of symbols which, when combined yield pictures intended to convey body language and facial expressions. This symbol system is so extensive that emoticon dictionaries have been published in order to guide newbies through this new language. The most basic emoticon is the 'smiley', which commonly appears as :) or :-). There are endless variations on smileys now, but some of the most commonly used are :(or :-(to represent sadness; :0 to represent shock or surprise; ;), the winking smiley, which represents sarcasm or an inside joke; :-> to suggest a devious smile; 8-) which is a smiley with glasses; the list is endless.

Beyond smileys, keyboard symbols are also combined to produce other representations of physical appearance or physical activity. For example, (@)(@) may be used to represent breasts; ( )*( ) may indicate that one has been 'mooned'; @>-,--'-- indicates that one has been offered a rose; and {{{{{{nick}}}}}} indicates that one has been hugged. As this communication system evolves, individuals are experimenting with altering the standard computer keyboard in order to allow for the development of new shapes and symbols to convey additional meanings.

In addition to the use of keyboard symbols, IRC participants may also use the "/action" command to simulate physical activity or indirect thought. In order to do this, the user types "/me stares at Fred" for example. To the other channel members, this statement appears on the screen as "Mary stares at Fred". Another possible use of the /action command is to simulate an indirect question or statement. This is done when one wants to make a point or raise an issue without violating the easy going feeling of IRC channels. So, for example, instead of asking "Fred, why did you say that?", which is in a threatening format that violates the friendliness norm, the statement would appear as "Mary wonders what Fred meant by that." In this way, one user may point out a potential problem to another without any sort of direct confrontation.

In general, emoticons and /action commands are developed and utilized in order to account for the most common emotional and physical requirements of a text-based identity. They allow for the efficient creation of a rich social context, in which users are able to convey socioemotional and physical cues. Such cues, in turn, allow for meaningful communication, the development of a sense of self, and a sense of group membership in an environment that might otherwise be devoid of meaning.

Finally, Undernet community members have enhanced the protocol of the server-server system to retard the occurrence of both lag and netsplit. The term 'lag' refers to the delay in messages reaching their destination. For example, if user number one does not receive any messages from user number two for several minutes, and then receives messages from user two all bunched together, user two is lagged. Or, if user one sees no messages from any user, and is then flooded with messages from everyone, user one is lagged. This can occur when users are not connected to servers that are closest to them or when a particular server is overloaded with users.

Netsplit, in a sense, is the ultimate form of lag. That is, when a server is severely overloaded, it can no longer accommodate all users and some users are 'split' from the 'real' channel off into what amounts to a parallel universe. They are separated from the real channel and cannot communicate with the members of that channel, even though the computer they are using tells them they are in that channel. Users must then sign off of IRC and sign back on in what they hope is the 'real' channel. However, this process can often take several minutes and as a result can be very frustrating.

It is easy to understand, given that IRC is a system of real-time chat, that both lag and netsplit effectively remove the affected individuals from the ongoing group conversation. The group conversation continues, and the affected individuals, for technical reasons, are prevented from engaging in it; one 'misses out' on what the group is doing until one can get back to the channel. Even when one returns successfully, that portion of the conversation that was missed is often gone forever, thus leaving the lagged or split individuals with holes in their understandings of the conversations. Particularly for the newbie, the experience of lag and netsplit can be frustrating because, not only does one not know how exactly to go about correcting the situation, one's attempts at 'fitting in' and becoming a member of the group are interrupted midstream.

Recognizing these issues as important in the development of stable channels, one of the primary goals of the individuals who split from EFNet to form the Undernet was to create server-server systems that minimized these possibilities. To that end, users have developed a protocol which disallows netsplit 'op riders'. If netsplit occurs in EFNet, it is difficult to maintain the notion that only one of the channels is the 'real' one. This is because people may intentionally induce a netsplit and then 'ride the split' into a channel and take over the status of channel operator, with all of its accompanying powers. Once in power, the new operators may do as they please, and the first thing they usually do is take the operator status away from the real channel operators. It is self-evident that such behavior, particularly when it is a constant possibility, makes the task of creating a sense of group cohesion and identity extremely difficult. Undernet protocol does not allow for such activities and, as a consequence, the stability of channel and participant identity is ensured. Even if some channel members are periodically lagged or split, there is the understanding that there is only one 'real' channel, the integrity of which cannot be violated.

These basic rules of netiquette, and systems of language, emoticons and action commands, combined with the technical improvements and the philosophy of those who started the Undernet, are what give the Undernet its feeling as a culture, as a place. And the feeling, or ambience, is that of Ray Oldenburg's [1991] 'third place'.

According to Oldenburg, the third place is "inclusively sociable, offering both the basis of community and the celebration of it. . . they (third places) host the regular, voluntary, informal and happily anticipated gatherings of individuals beyond the realms of home and work." [p. 16] He goes on to describe the eight basic characteristics of third places as follows: 1) they are established on 'neutral ground'; 2) they are social status levelers; 3) conversation is the main activity; 4) they are accessible and accommodating in terms of time and location; 5) they are given character by 'the regulars'; 6) they keep a 'low profile'; 7) the mood is a playful one; and 8) the environment is congenial--a home away from home.

As any Undernet channel regulars will attest, IRC is a prime example of a third place. IRC channels are neutral ground; participants can come and go as they please, and no one has to be the host (though everyone feels at home). Channels are also status levelers for participants. That is, they are accessible to the general public, do not set formal criteria of membership or exclusion, expand the types of people one can get to know, and put the emphasis on personal qualities not confined to status distinctions. While this does not necessarily mean that everyone is perfectly equal in the Undernet, what it does mean is that "the charm and flavor of one's personality is what counts. . .Those who, on the outside, command deference and attention by sheer weight of their position find themselves in the third place enjoined, embraced, accepted and enjoyed where conventional status counts for little." [Oldenburg, p. 24]

Furthermore, it is self-evident that conversation is the main activity of the Undernet; conversation is all there is. But IRC's quality as a third place goes deeper than this. Not only is conversation the main activity, it is the type of conversation which is important. It is lively, scintillating, colorful and engaging. IRC dialogue conforms to the 'simple rules of good conversation': people share the talking time relatively equally; they are attentive while others are talking; they are careful not to hurt others' feelings while still managing to say what they think; they avoid topics not of general interest; they say little about themselves personally (at least during public chat room conversations); they avoid trying to instruct others (instead making suggestions about possible issues); and they speak in 'as low a voice' that will allow others to 'hear'. All people contribute about the right amount of time, and most everyone participates to some degree, even the newbies.

Undernet channels are 'open' 24 hours a day, and one may go alone at almost any time of the day or night and see people one knows. And, while there exist groups of regulars at various points during the day, there is fluidity in the timing of meetings--people come and go at somewhat similar hours, but not exactly the same, and sometimes days are missed by some regulars. Of course, these days missed are duly noted and commented upon by the regulars who do drop by; people are questioned as to the whereabouts of someone who is missing. Activity is unplanned, unscheduled and unstructured, and that is what lends the group character.

And what of the regulars? Oldenburg suggests that it is the regulars that gives the third place its conviviality: "Third places are dominated by their regulars, but not necessarily in a numerical sense. It is the regulars whose mood and manner provide the infectious and contagious style of interaction and whose acceptance of new faces is crucial." [p. 34] In IRC, just as in other third places, one becomes 'a regular' through a certain process. Acceptance into the circle is not difficult, but it is not automatic, either. A newbie must establish trust between him- or herself and the regulars; one continues to reappear and tries not to be obnoxious. No 'serious' conversation is allowed; personal problems and moodiness must be set aside. Conversation is lively, and consciousness of time and circumstances slips away. Ultimately, "every topic and speaker is a potential trapeze for the exercise and display of wit. The unmistakable mark of acceptance into the company of third place regulars is not that of being taken seriously, but that of being included in the play forms of their association." [p. 38]

Third places are said to have a low profile physically; often they were originally designed to serve a different purpose, and haphazardly evolved into places for sociability. Clearly, IRC channels represent the lowest 'physical' profile possible, as they are socially created places. For its members, IRC is a 'home away from home'. It roots people in time and place as people tend to go at around the same time everyday and see familiar faces. It allows for a sense of control over one's environment as regulars are accorded special privileges not granted to others. After visiting one's channel, one feels socially regenerated and restored.

Thinking back to much of the anecdotal critique of CMC, it tends to revolve around the idea that such forms of communication are an 'escape from reality'. Oldenburg argues that, while third places are an escape, they are not just an escape. To view such places in this manner emphasizes the outside world and neglects the examination of what happens inside third places.

THE USENET COMMUNITY

While no one would argue that Usenet is a third place, it still displays all of the basic attributes of a community. Usenet most definitely has its own system of language, as well as a clearly defined set of beliefs, values and norms. The difference is ultimately one of style.

According to Tom Seidenberg, the best way to describe Usenet culture is through the 'room analogy':

> "You've got this enormous building with thousands of rooms. Each room has a sign outside describing what is being discussed inside. . .Some rooms are very organized. These rooms have a large audience and a few selected speakers, and a spokesman or two. . Some rooms do not have any spokesmen, but are still very well organized. Every now and then some people go over to a corner and have a quiet conversation, but nothing terribly loud. Other rooms are like big social gatherings with many smaller groups talking (or yelling) among themselves. There are many people in these rooms, most of them just walking around and listening in on the various conversations. . .There are also play rooms, some with very few people, but they make a terrible racket! Often, these play rooms have a king of the mountain, a demi-god, or even a bully. . . "[1995, pp. 3 & 4]

This is an excellent explanation of Usenet culture because it simultaneously expresses the idea that each individual Usenet newsgroup has its own particular atmosphere, while at the same time all newsgroups share certain basic values about how Usenet life should be conducted.

There are two seemingly contradictory beliefs that participants in Usenet hold about their community, both of which are claimed to stem from consensus of personal experience. The first generally held belief is that Usenet is anarchy. The second belief is that Usenet behavior is fairly predictable. For example, according to Henry Hardy:

". . .Usenet is unusual among computer networks or communications systems in that it has no formal rules, no formal enforcement mechanism. . . .The combination of public, nonprofit and commercial networks over which Usenet is transmitted makes for a complex and controversial situation with regards to liability, free speech, obscenity, appropriate use, commercial use and other issues. The lack of a central regulatory or governing body makes the Usenet a study in functional anarchy. . .The current decision making structure of usenet might best be described as 'cooperative anarchy'. Usenet operates more like a culture than a formal system. The rules of conduct are implicit and are only explicitly stated in the news.* hierarchy groups and in some FAQs. The FAQs and the memories of the 'ancient ones' represent the collective wisdom of the net." [1993, pp. 2, 4, & 7]

Such descriptions of 'Usenet as anarchy' are fairly common in much of the written documentation on the history and culture of this community. The basic premise is that, because there is no single central governing body, Usenet is an anarchic system. However, the cultural beliefs outlined below regarding the predictability of Usenet social life belie the proposition that Usenet is anarchy. In fact, a lack of central, formalized controls have lead to the development of sophisticated, if informal, means of social organizing; means that are 'formal enough' to accomplish the tasks at hand.

Regardless of the fact that the Usenet community enjoys and perpetuates the myth of anarchy, 'true' members of the Usenet community believe that they are capable of predicting Usenet social life. This belief is so strong that it has been formulated into a set of laws and concepts which are generally understood by the core Usenet population.

Among the most important laws and concepts are: 1) Aahz's Law--the best way to get information on Usenet is not to ask a question, but to post the wrong information; 2) Boigy's Law--the theory that there are certain topics in every newsgroup that are discussed cyclically. Often, the period of the cycle and the length of the resulting discussion, can be accurately estimated by those who have been around long enough; 3) Sturgeon's Law--ninety percent of everything on Usenet is crud. What that ninety percent is depends on who you are. This does not imply that the remaining ten percent is not crud; 4) Religious Issues--questions which seemingly cannot be raised without touching off Holy Wars. Almost every group has a Religious Issue of some kind, and they are guaranteed to start a long running, tiresome, bandwidth-wasting flame war; 5) September--the time when college students return and start to post stupid questions, break rules of netiquette and just generally make life on Usenet more difficult. However, with the growing popular media discussion of the 'information superhighway', it now seems like September lasts year round; 6) Signal to Noise Ratio--a subjective quantity describing someone's idea of just how much content a group has, relative to the junk the group has. Each person has their own threshold and if a group falls below that threshold, a member will unsubscribe; and 7) The Imminent Death of Usenet--a cyclical prediction which shows up in the popular press and references the assumption that Usenet is anarchy; the latest upheaval, technological or social, is more than likely to bring about the demise of the Usenet system. This particular term is now a myth-become-joke among core Usenet members. [alt.culture.usenet FAQ 1995, 5- 6]

Beyond typifying laws of behavior, Usenet participants also classify some personality types within their community. For example, there is the Kook--a weirdo who randomly appears in random groups and who causes no end of trouble. Often placed in kill files[3], the kook usually only disappears after everyone finally stops responding to their ridiculous or improperly posted articles. Then there is the Lurker--one who reads a newsgroup but posts no articles. The Lurker has reached such mythical proportions as to provoke estimates that 90 percent of people who use Usenet are lurkers.

And finally, there is the Net.God, which is a concept best understood through use of a specific example. According to the Net.Legends FAQ [1995], Kibo is a genuine net.god. "Perhaps the single defining hallmark of genius is to do something that no one else has ever done before, or even thought of doing, and making it look blindingly obvious afterwards. James 'Kibo' Parry, confronted with the vast reality of Usenet, decided to 'grep' his entire newsfeed for posts containing "kibo" in order to look them over and see if they were worth replying to. He has now become a Usenet term: grepping your newsfeed is "kibozing", and one who does so is a "kibozer". He also has his own newsgroup: alt.religion.kibology, which is also his own religion. If you receive email from him, you are given a kibo.number."

This typification of social phenomena and personalities in Usenet has become extensive enough to warrant a document (cited above) known as the Net.Legends FAQ. The Net.Legends FAQ is described as ". . .a collection of descriptions of net.phenomena one hears about in passing and wishes one had more information about. Not all are completely factual entries: in some cases the true facts are known only to one person, or lost in the mists of time, while in others the facts pale in relation to the mythology. In any case, the actual facts included, sparse though they may be, are true as far as I know. Since these are net.legends, not much of a real attempt to verify this information has been made." Despite the description, within the Usenet community this document represents 'the truth' with relation to social history. This is the case with all FAQ documents; they establish Usenet as a community. And, while they may not be 'formal' means of social control, when referenced they carry a great deal of weight in making a case for or against certain actions or behaviors.

One can see that, sociologically, there is a "we" in Usenet and this "we" has a history together. Further, "we" refer to ourselves as being present in physical places--rooms. We share beliefs about ourselves and our social system and we also believe that we know ourselves, each other and our relation to the 'outside world' well enough to predict the behavior of those around us. In the face of evidence to the contrary and self-identification as a community, Usenet members continue the myth of anarchy. It is a myth that allows for the development of group identity relative to the outside world; a world which is, to a large extent, complicit in perpetuating the myth of anarchy. This belief allows members to feel a sense of solidarity and of participation in the development of a new form of social organization; something which has never been done before. Of course, this myth of anarchy is predicated on the assumption that anarchy is the opposite of order. As Robert Bierstedt [1970] points out, however, the opposite of culture or order is not conflict or anarchy or disorder. The opposite of order is non-involvement.

Like IRC, the basic values in Usenet are efficiency (relative to time and bandwidth), and quality communication. Accordingly, participants in Usenet have developed a system of netiquette and an interrelated, Usenet-specific language-form (the troll), which are intended as informal means of maintaining these core values. There are several documents that lay out Usenet rules of behavior. Some of them are designed for users, and some of them are designed for system operators (sysops). However, they all emphasize two main points: conservation of scarce resources, and being generally 'polite' to others, thus creating a positive self image.

For example, according to Chuq von Rospach's *A Primer on How to Work With the Usenet Community*, users are encouraged to: 1) never forget that the person on the other side is human; 2) be careful what you say about others; 3) be brief; 4) remember your postings reflect upon you-be proud of them; 5) think about your audience; 6) be careful with humor and sarcasm; 7) only post a message once; 8) rotate (encrypt) material with questionable content, answers or spoilers; 9) summarize what you are following up; 10) use email-don't post a follow-up; 11) read all follow-ups and don't repeat what has already been said; 12) double-check follow-up newsgroups and distributions; 13) be careful about copyrights and licenses; 14) cite appropriate references; 15) do not ignite a spelling flame war; 16) don't overdo signatures; 17) do not use Usenet as a resource for homework assignments; 18) do not use Usenet as an advertising medium; and 19) avoid posting to multiple newsgroups. Clearly, two key issues are concern for other users' feelings and impressions of you and concern for the conservation of scarce Usenet resources.

Of course, in order to have a full appreciation of Usenet etiquette, the role it plays in establishing social order and the degree of seriousness community members attach to those rules, one must read them through the satirical framework that gets to the heart of the 'feeling' of Usenet; one must read Brad Templeton's Dear Emily Postnews. Consider, for example, netiquette rule number four--'your postings reflect upon you, so be proud of them'. Within the Emily Postnews format, this issue is addressed as follows:

Question:
"Dear Emily Postnews,
I can't spell worth a dam. I hope your going to tell me what to do?
Answer:
Don't worry about how your articles look. Remember it's the message that counts, not the way it's presented. Ignore the fact that sloppy spelling in a purely written format sends out the same silent messages that soiled clothing would when addressing an audience."

Or, consider netiquette rule number ten--use email, don't post a follow-up:
Question:
"Dear Emily Postnews,

I read an article that said, 'reply by mail, I'll summarize'. What should I do?

Answer:

Post your response to the whole net. That request applies only to dumb people who don't have something interesting to say. Your postings are much more worthwhile than other people's, so it would be a waste to reply by mail."

Ultimately, in the Usenet community, a user's postings are his or her identity. And a critical point in defining oneself as a good community member is recognizing that one's time and thoughts are no more (or less) valuable than those of other users. Usenet is not intended as a one-way broadcast medium. It is for people to carry on discussions about specific topics in which they are already interested. Using Usenet in a means which violates this rule, commonly called net-abuse, is a serious offense.

Besides using rules of netiquette as a means of maintaining values and conveying meaning, Usenetters utilize both the language of the emoticon (as did IRC members) and Usenet specific language forms designed for the control of 'deviant' behavior. There has been a great deal of confusion about two Usenet language forms; the flame and the troll. Briefly, to flame in Usenet means "to post a message intended to insult or provoke; to speak incessantly and/or rabidly on some relatively uninteresting subject or with a patently ridiculous attitude; either of these two, directed with hostility at a particular person or persons, for example, posting an article on how to run over dogs in rec.pets.dogs." [alt.culture.usenet FAQ 1995, 5] A troll, on the other hand, is a posting that is designed specifically to generate followups about something trivial, but not in the sense of a flame, or a post designed to instruct readers to ignore obvious drivel by making those who reply feel utterly stupid. Such things as misspellings, incorrect facts or concepts can be used as trolls. After the followups have died down, the troller will usually inform the victims of their status as guinea pigs and then move on to another group.

Now, the reason that it is critical to bring up this particular 'deviant' behavior here is that the argument against Usenet as 'real' community rests in large part on the notion that interaction in Usenet is 'uncivilized'. It is claimed that in Usenet, it is too easy to flame people and discussions often break down into name-calling that would not be tolerated in a real-life meeting or social setting. This claim demonstrates a lack of understanding of Usenet culture.

First, one must consider the context of the flame. While it can be, and often is used as described above, simply to 'irk' people, it is also used in boundary work. That is, as one experienced Usenetter says, "we especially like to flame people who even hint at newbieness." Thus, flames have multiple purposes, one of which is to make the social distinction between the 'lowly' newbie and the 'true' Usenetter; being flamed as a newbie is a rite of passage.

Second, when the flame does consist of a 'patently ridiculous attitude', its use is controlled by community members, sometimes through use of the troll:

> "The purpose of trolling is not to deliberately start a flame war. Deliberately starting a flame war is flame-baiting, and it requires absolutely no intelligence to post something obscene that'll get people mad. Trolling is more subtle. It's a tactic to discourage flaming, by posting intelligently and cleverly crafted (but marked) inaccuracies; someone attempting to flame a posted troll finds that he has acted too rashly and has succeeded only in making a fool of himself. It reinforces netiquette; flamebaiting encourages the breakdown of netiquette. A good troll is an impressive thing." [alt.culture.usenet FAQ 1995, 8]

Though it is markedly different from IRC, Usenet may easily be viewed as a community with its own distinct culture. It displays norms, customs and traditions different from the external cultures in which it is embedded. Enforcement of these norms begins with the individual community members, and consensual interpretation by the Usenet public becomes the 'law'.

THE FIDONET COMMUNITY

Fidonet is no longer just a piece of software;
it has become a complex organism.

Tom Jennings, Fidonet founder

The easiest way to get a general sense of Fidonet culture is to draw some comparisons between it and Usenet culture; both are asynchronous messaging systems which are similar enough in technical structure and purpose that the main concerns of Usenet (conservation of bandwidth, efficiency and proper presentation of self) are also the main concerns of Fidonet. However, where Usenetters see their community as "anarchy" (even though behavior is considered predictable), the basic belief system within Fidonet is that volunteerism and cooperation combine to create an orderly (enough) community.

The primary difference between Fidonet and Usenet is one of funding. While Usenet is funded largely through universities and commercial sponsors, Fidonet is a completely voluntary system. Further, while Usenet was designed with 'serious' concerns in mind (e.g. transmission of data among universities), Fidonet's purpose, "...very simply, is as a hobby; a non-commercial network of computer hobbyists ('hackers' in the older, original meaning) who want to play with, and find uses for, packet-switch networking. It is not a commercial venture in any way; Fidonet is totally supported by its users and sysops (system operators), and in many ways is similar to ham radio, in that other than a few 'stiff' rules, each sysop runs their system in any way they please, for any reason they want." [Jennings 1985, 3]

Fidonet was created, by accident, in June,1984 and had only two nodes. By August of that year, it had 30 nodes. This pattern of growth quickly became the justification for a new system of message routing. While this routing system required additional people to assist with the network, it remained a cooperative voluntary system; it did not immediately become a bureaucracy. According to Jennings, "Well, at first, everyone knew each other; we were in more or less constant contact. However, when the node number got into the twenties, there were people bringing up FidoNodes who none of us knew. This was good, but it meant that we were not in close contact anymore. The Net started to deteriorate; every single week without fail there was at least one wrong number, usually two. Fidonet is just too large today to run as an informal club". [1985, p. 2] This process of routing messages in a low cost manner to all nodes was how the zone/region/net/node/point hierarchy explained in Chapter 2 came about; the original intent was to "decentralize Fidonet" and maintain messaging efficiency.

However, Randy Bush was correct in his assessment that, because Fidonet has been operated by end-users and hobbyists, rather than 'computer professionals', it has experienced social and political issues more rapidly and more seriously than have other network cultures:

"In 1986, a well-intentioned but naive group formed the International Fidonet Association, intending to promulgate the technology and coordinate publication of the newsletter (FidoNews) and other writings about the network. Unfortunately, as Fidonet operators were far more socially oriented than their more technical brethren in the other networks, the formal organization of IFNA tended to draw considerable political interest and attracted the less constructive political elements of the Fidonet culture. The issue came to a head in 1989 with an attempt to load the IFNA board of directors and pass a motion which explicitly put IFNA in complete control of the network. The motion was cleverly forced into a netwide referendum (Fidonet's only global vote to date) which required a majority of the network assent to IFNA rule. The referendum did not pass, and IFNA was subsequently dissolved." [1993, p. 7]

In 1987, what is known as *Policy4* came into existence. The first policy document of Fidonet, Policy1, was written in 1985 and adopted by informal consent. "It was mainly a record of currently accepted policies. It worked for a long time and only became obsolete as we grew in complexity. It was good. This original policy grew and grew, and somehow culminated into the monstrosity of *Policy4*, which is not a policy, but an attempt at law, related to Policy1 only in name. It stinks." [Jennings 1992, vol. 9-34, 1]

Policy4, written by Regional Coordinators, has 'a large amount of social and political content enshrining a hierarchy of coordinators', all of whom are either elected or appointed by those Regional Coordinators. Beyond that, and perhaps even more importantly, the *Policy4* document also has a 'tone' to it that violates Fidonetters' commitment to voluntary cooperation. The tone violates basic cultural beliefs by assuming that Fidonet community sysops, to whom it is directed, need some sort of official document that will tell them, in detail, every responsibility and obligation they have in Fidonet life; that without such dictates, "Fidonet is large enough that it would quickly fall apart of its own weight unless some sort of structure and control were imposed on it." [p. 1] It is a legally stylized document in its numbering system, and many sentences begin with 'You are responsible for...', 'It is your responsibility to...', 'You must...', or 'You should...'. One Fidonet member, to point out the offensiveness of such a tone and the ridiculousness of the formal hierarchy, wrote a satire of *Policy4* which culminated in, "These levels act to put everybody under the thumb of whoever takes charge; this is considered desirable because the author of this document is in charge." [Priven 1988, vol. 5-06, 10]

There have been other political upheavals in Fidonet, for example the Zone 2 War during 1992. Such problems have caused enough concern within the Fidonet community that the 'Fidonet History Project' was begun in 1994. The Fidonet History Project is an attempt to chronicle the first ten years of Fidonet's growth. It was begun because there was concern that,

> "...with all of the changes that have occurred in just the last year or two, Fidonet is in very real danger of losing its roots. We have a noble heritage being the first grassroots communications network, literally by the people and for the people. It would be a shame to lose that. Fidonet is a rather unique entity that tends to defy description... it has burgeoned into an indescribably loud and boisterous while at the same time quiet and thoughtful communications network comprising sysops and users from all walks of life. I greatly fear that no writer...can do justice to Fidonet, its culture and tradition. Yet I feel it's important that Fidonet's roots be preserved for future generations of cybernauts." [Robbins 1995]

The main point to be gleaned from this brief overview is that Fidonet has a history and tradition that its members are proud of. As with both Undernet and Usenet, one definitely gets the sense from this description that Fidonetters have a sense of community. Not only do they get into heated political debates about where they are going as a group, they have set up an organization to preserve their cultural heritage. And, as with the other CMC communities previously discussed, they talk about themselves as being co-present in physical places, and they design norms of behavior, systems of language, and formats for communication (specifically, FidoNews) to endow those physical places with meaning. Consider the following description of Fidonet by founder, Tom Jennings [1985]:

> "In the dark ages of modems (pre-1982), there were so few bulletin boards and users that there basically wasn't a problem. You somehow managed to get a modem . . . and started dialing. You got nervous and made a mess of the message base, and if you were real

unlucky, crashed the board. Everyone knew you were 'new', and so were tolerant while you learned how to get around. Crashers and trashers weren't really a problem. . . .Now, however, it frequently becomes a situation like a traveler to a foreign country who is totally unfamiliar with local customs. Visitors embarrass themselves by saying the wrong thing, or insult the locals with totally inappropriate reactions. . .In face-to-face encounters with people that you don't know well, there are thousands of unwritten rules that just about everyone follows. A big problem with modeming is that you miss all non-verbal communication details. You have to make up for this in other ways. Get an idea of what kind of people are there. Bulletin boards are no different than a local bar or whatever. . . .This is no different than joining a conversation at a party or cafe; you just can't jump in and blaze away with your wit. . . .Keep in mind that some things that are wonderful person to person can be absolute disasters in print." [pp. 2,3]

Bearing in mind this sense of place, combined with the need for efficient, low-cost communications, Fidonetters have developed two simple rules of netiquette (which parallel the rules of IRC and Usenet): 1) thou shalt not excessively annoy others; and 2) thou shalt not be too easily annoyed. In order to avoid 'excessively annoying others', one should: 1) remember that each conference in the bbs has a subject--don't get too far off of it; 2) be cautious about starting a new topic if one has newbie status; 3) in responding to a message, mention enough of the previous message so that other readers can tell what you are replying to; 4) remember the delay involved in Fidonet messaging; many may be replying to a piece of mail at once, and you may get repeats or a flood of similar responses all at once; 5) keep messages short; and 6) avoid flaming; it is bad form to spend more time in the reply discussing personalities than the real issues.

In order to avoid 'becoming excessively annoyed', one should: 1) remember that people may call in everyday, once a week or maybe never again; be patient when waiting for a reply. After a while, one gets an idea of who calls in how often; 2) remember that, in the user-sysop relationship, "There is a human out there somewhere. Sysops are saints and assholes like everyone else, and they have the responsibility of keeping the system up and running. Getting angry at the sysop for not answering a request as quickly or as thoroughly as you want is definitely not the way to get in good graces on that board." [Jennings 1985, vol. 2-30, 4]

Given the previous discussion of emoticons and their ability to enhance meaningful communications, it is unnecessary to discuss this again for the Fidonet system. Suffice it to say that Fidonetters use emoticons in the same way Usenetters do, and like Usenetters, they do not have as extensive an array of emoticons as do Undernetters. Like Usenetters, they tend to use emoticons relatively infrequently, and to use only the more basic styles of smileys. However, Fidonet does employ one additional communication format that is critical to the establishment and maintenance of community; the weekly newsletter, FidoNews.

Like the FAQ system in Usenet, the FidoNews newsletter brings together what might otherwise be isolated and individualized bulletin board systems (BBSs). Like Usenet newsgroups, each Fidonet BBS has a somewhat different 'feel' to it. Some are centered around playing games, while others may be family-oriented. But, as the FAQ documents in Usenet provide all newsgroups with a sense of belonging to something larger, so does FidoNews for BBSs. This fact is readily acknowledged by Fidonet founder, Tom Jennings, and by others in the Fidonet community. "The newsletter, FidoNews, was and still is an integral part of the process of Fidonet. FidoNews is the only thing that unites all Fidonet sysops consistently." [Jennings 1985, 8]

FidoNews provides a sense of being a community of people with common interests; it provides, what Benedict Anderson calls the imagined community:

> "We know that particular morning and evening editions will overwhelmingly be consumed between this hour and that, only on this day, not that. . .this mass ceremony is performed in silent privacy, in the lair of the skull. Yet each communicant is well aware that the ceremony he performs is being replicated simultaneously by thousands of others of whose existence he is confident, yet of whose identity he has not the slightest notion. . . .What more vivid figure for the secular, historically clocked, imagined community can be envisioned?" [1991, p. 35]

While FidoNews is not necessarily read simultaneously by all sysops, it is a regular, weekly ceremony, performed in privacy, through which individuals develop the sense that they know other Fidonetters and belong to that community. As founder Tom Jennings points out: "There are quite a lot of us here. About a thousand sysops, and at least ten times that many users. Many of us have gotten to know each other quite well. This is quite amazing, especially when you consider that few of us have ever met. We know each other by the words we type. We see each other as little dots of light forming text on our screens." [1986, vol. 3-23, 2]

ENDNOTES

[1] Culture is generally defined as a shared way of life; a complete social heritage. It represents the beliefs, values, behaviors and material objects shared by a particular people. The major components of culture are: 1) symbol--anything that carries a particular meaning recognized by members of a culture; 2) language--a system of symbols that allows members of a society to communicate with one another; 3) values--standards by which members of a culture distinguish the desirable from the undesirable, what is good from what is bad, the beautiful from the ugly; and 4) norms-- rules that guide behavior

[2] 'Bot' is short for robot. There are many types of bots on IRC, some of which are necessary and helpful to channel operators. However, clonebots and floodbots serve only to harass others and clog IRC channels, often resulting in 'lag' and 'netsplit', which are two serious problems among IRC users; lag and netsplit prevent the easy, ongoing and lighthearted conversation which is the raison d'etre of IRC.

[3] A 'kill-file' is a program that users can install in their newsreaders that will automatically delete unwanted postings from the selected individual. This way, the user is not required to sift through what he or she considers to be waste-of-time postings.

CHAPTER 4

UNDERSTANDING COMMUNITY STRUCTURE: SOCIAL INSTITUTIONS AND GROUPS

In general, human communities must solve a few basic problems in order to survive. They must establish a system of values and norms of behavior that members live by. They must socialize new members effectively so that they, too, abide by these rules. They must develop a system of sanctions that can be used to hold individuals accountable for their actions. And they must provide the means for individuals to simultaneously develop distinct identities and a sense of group membership. Thus, when attempting to understand a new culture, a good place to start is with an examination of the formal and informal strategies the group uses to achieve those goals. What kinds of institutions and organizations have people created to solve these problems of human group life? Are these strategies effective in meeting the needs of the group while still allowing for personal identity?

In trying to answer such questions, it is common practice for professional sociologists and for casual social observers alike to assume that the primary means by which human communities achieve these goals is through the creation of formalized organizations. Social groups are generally classified as primary or secondary. According to Cooley [1913] primary groups are those characterized by intimate face-to-face association and cooperation. They are fundamental in forming the social nature and ideals of the individual. These groups, for example, the "family", are the essential mechanism of socialization and the primary source of social order; they represent a sense of 'we' in that they involve the sort of sympathy and mutual identification for which 'we' is the natural expression.

When such a group makes patterns of interaction explicit through writing down rules, it becomes a secondary group, or formal organization. As a small group increases in size, it reaches an upper limit at which point the group is altered and establishes 'formal' rules and regulations. In the formal organization, relationships are impersonal, everything is written down, and members rely on rules that can be read. However, it is also noted that:

> ". . .formal organizations inevitably inspire the formation of informal patterns which often become more important than the formal. The way things are supposed to be on paper is balanced by patterns that actors negotiate on their own in face-to-face interaction. Formality aids people when they interact--it makes it relatively easy for new members to know very quickly what to do--but it is usually more important to alter the written patterns and bend them to fit our own situation, since those who wrote them could not possibly know our situation exactly. Indeed, most people in formal organizations seem to understand that written rules are guides which are not usually strictly adhered to." [Charon 1993, 50]

However, this distinction, upon which many of the critiques of CMC are implicitly based, discounts the importance of many, more 'informal' means of social organization and problem-solving. Means which all community members are capable of using on a daily basis to maintain social order and group cohesion. Regardless of whether these distinctions are valid for face-to-face communities, they detract from one's ability to understand on-line communities for two reasons. First, these definitions rely on the idea of 'face-to-face' interaction as the standard by which all other communication is to be judged. CMC problematizes this standard because it de-centers the notion of place; it cannot be said with any certainty whether any type of CMC interaction is face-to-face or whether it is not. This whole notion relies on physical proximity and, as such, is irrelevant to the understanding of CMC groups.

Second, the assumption that written rules are a determining factor in whether or not a given social arrangement is formal or informal is clearly problematic. In a text-based reality, everything is written down one way or another; text is all there is. One is forced to ask the question, "how can 'written down' be used as a standard of formality?", and the only conclusion to be reached is that participants in CMC behave as if they have 'official' policy. It is this belief in and willingness to obey rules that makes those rules 'real'. Claiming that on-line groups do not constitute communities based upon these standards is circumspect.

A more fruitful way of looking at social groups, including on-line groups, is through Don Martindale's [1962, 1966] framework. According to Martindale, "The human group is an organized pattern of interhuman activity; it is not a collection of individuals. . . .There are two fundamental reasons why the individual should not be considered the fundamental unit of the group. First. . . no group exhausts all possible behaviors of the individual. . . .The second reason. . .is that within remarkably wide limits, one individual's activity can be substituted for another's, while the group remains essentially unchanged. The true unit of the group is not the individual, but the activity or performance." [p. 120]

He goes on to argue that organizations, like any human group, are strategies of collective life. Defining human groups as concrete patterns of interactions, rather than as collections of individuals, eliminates the need to distinguish between groups which are 'formal' versus 'informal'. What needs to be understood are the various types of interaction and how they are utilized in order to establish and maintain social order and social identities for the individuals involved. There is nothing 'magical' about the presence or absence of face-to-face interaction or written rules when one is attempting to discern the presence or absence of social order. What is important is whether participants in CMC act as if the means of maintaining social control exist and behave accordingly; do they agree to agree.

Similarly, social institutions may be thought of as organized patterns of behavior. Rather than being concrete social structures, they are ultimately problem-solving strategies. Drawing on the work of Cooley [1913], Hertzler suggests that, "Institutions are first and foremost psychic phenomena. The institution has primarily a conceptual and abstract, rather than a perceptual and concrete existence. Their essence is ideas and other concepts, interests, attitudes, traditions. In a very real sense, institutions are only in our heads; they are common and reciprocating attitudes." [p. 35] He notes, however, that institutions are also 'societal structures', secondarily, in the sense that various social groups with norms of behavior and 'physical extensions' utilize these psychic phenomena to create and justify systems of social organization.

Don Martindale [1962, 1966] continues this characterization of the relationship between the institution and the group by arguing that institutions are the relational patterns manifested by groups. Groups are organized sets of interhuman behaviors, and institutions are the relations displayed therein:

"Groups are systems or patterns of social behavior which arise when pluralities pursue their individual and collective aims in common. . . Groups are not something different from social behavior; they are merely special semi-established regularities of social behavior. Institutions are defined as the standardized solutions to the problems of collective life. . . A group is a strategy of interhuman behavior, that is, a plan of action intended to achieve common objectives. . . .A group is a concrete system of activities; a group institution is the solution to the problems of social life." [pp. 39-46]

For example, "the family" is both a social institution and a group. The concept of 'family' is an institution in the sense that it solves the collective problem of socialization. It is an ideal, a concept, that most community members value. Though its concrete form may change over time, the idea of family remains relatively constant. In a more concrete sense, 'family', one's own family, is an actual social group with regularized patterns of behavior.

What does all of this mean for CMC communities? It means that in order to understand life on-line, one must set aside unexamined assumptions and allow oneself the opportunity to learn about the systems of values, norms of behavior, and the problem-solving strategies that on-line groups have devised within the contexts of the techno-social environments they confront.

THE INTERNET RELAY CHAT COMMUNITY

Internet Relay Chat members have developed several institutional strategies for dealing with community problems. Two of these may be thought of as similar to 'real life' government structures, while the others are more closely related to the day-to-day management of life in individual IRC groups.

The two basic IRC groups that deal with broad technical and social issues are the Boards of IRC Coordinators, or BICs (for all of IRC) and the Undernet User's Committee and Group (for the Undernet only). The BICs (for example, EBIC in Europe and USBIC in the United States) set guidelines for technical issues, including 1) establishment of new servers; 2) linking of servers; 3) 'cracked' servers; and 4) routing plans for server backbones. Additionally, these groups also deal with social issues of IRC, including: voting issues, the appointing of new IRC operators and penalties for rules violations among IRC operators.

The Undernet User's Committee, according to coordinator Warren Aluve [1995], "consists of a small group of concerned and interested users. It's purpose is to inform users of developments that may interest/affect them, communicate user proposals/views to other committees and the caucus, and ensure smooth running of the User Group as a whole. The User Committee shall not mediate disputes between users and channel ops or arbitrarily decide on an issue, unless it directly affects the operation of the User Committee." [p. 2] The committee may hold meetings when necessary and vote on issues, may create working parties drawn from the User Group to perform certain tasks or work on projects.

The User Group refers to the mailing list as a whole. It is the forum for the proposal and discussion of ideas, the formulation of methods to achieve those ideas, voicing of concerns, keeping the users informed of developments, and avoiding and solving oper/user disputes which may arise (meaning IRC opers, not channel ops). Like the User Committee, the User Group is not for: 1) disputes between users; 2) discussion of the politics of any channel; or 3) resolving conflicts which may arise between channel ops and users of the Undernet. [Undernet User Committee Guidelines 1995]

The significance of these groups is that they provide, through reasonably democratic means, a framework of operations for IRC as a whole. In doing so, however, they set out to be deliberately non-authoritarian, providing only 'guidelines' for technical and social issues. While they serve as a means through which individual members of IRC channels may define themselves as part of a larger 'whole', within Undernet, there are several more immediate institutions and groups that allow members to maintain social order and individual identities.

A major threat to social order in IRC is the ever-present potential for netsplit and the resulting channel wars. The people who created the Undernet specifically designed a code system and a bot, "X", to eliminate these problems. According to an anonymous but reliable source:[1]

> "Undernet has a more solid system code, which does not allow people to ride on a netsplit. You know that in a netsplit, the net gets divided into two parts. Depending on where the physical location is of

the net disruption (like where on earth), sometimes one can have only 10-20 people, while the rest (can be thousands) are on the other side. Here is the trick!! (Shhhh...don't tell anyone hehe...this is how you take over a channel *grin). You wait for the split, knowing that every now and then a net split occurs. Say you are on channel #C, once you are in the middle of a split, there co-exists TWO channels with the name #C. You act fast, oping yourself (assuming you are alone on channel #C at your side of the split. You achieve this by choosing a small server, with few users), and then just wait for the people on the other #C to come back from the split. As soon as the split is over, everyone starts coming back, and there will be only one #C again. You must be fast here! As soon as the first person comes on #C from the other side you de-op him! *grin*..and you de-op the next, and the next...If you don't -- they will de-op you...Now you are the only ops on the channel, and if you don't have any scruples, you can feel proud of yourself...The people who were ops before the split will of course protest, and if some of them are armed with a war script..it's gonna be a big battle starting..."

For the individual caught in a netsplit, the following is a typical example of what appears on the computer screen. Netsplit is usually preceded by 'lag'. Initially, the server randomly bans people (mode = +b) from channel #25plus, then allows them to rejoin. This is followed by a brief discussion of the split, which can occur repeatedly and last for extended periods of time:

[DISP-#25plus][ACTION]   Marcia_Br is suffering major lag. brb
[DISP-#25plus][LOCAL]   Leaving ...
[DISP-#25plus][LOCAL]   CHANNEL LOG ON
[DISP-#25plus][LOCAL]   WAR MODE OFF
[DISP-#25plus][LOCAL]   FAKE OPS ON
[DISP-#25plus][LOCAL]   Refreshing PROTECT and OPERATOR LISTS
[DISP-#25plus][LOCAL]   Done Refreshing
[DISP-#25plus][SERVER]   PaulaV!~pvcrazy@ppp162.iadfw.net has joined this channel
[DISP-#25plus][Manhattan.KS.US.Undernet.Org-MODE]   Has changed #25plus's mode to +nt

Understanding Community Structure: Social Institutions and Groups 51

[DISP-#25plus][Manhattan.KS.US.Undernet.Org-MODE] Has changed *!*alexb@*.pcdocs.com's mode to +b
[DISP-#25plus][Manhattan.KS.US.Undernet.Org-MODE] Has changed HOTGUY!scoto@ix-ftl2-17.ix.netcom.com's mode to +b
[DISP-#25plus][Manhattan.KS.US.Undernet.Org-MODE] Has changed *!*John@*.sc.net's mode to +b
[DISP-#25plus][Manhattan.KS.US.Undernet.Org-MODE] Has changed jwheeler!James@buffnet1.buffnet.net's mode to +b
[DISP-#25plus][LOCAL]   Leaving ...
[DISP-#25plus][LOCAL]   CHANNEL LOG ON
[DISP-#25plus][LOCAL]   WAR MODE OFF
[DISP-#25plus][LOCAL]   FAKE OPS ON
[DISP-#25plus][LOCAL]   Refreshing PROTECT and OPERATOR LISTS
[DISP-#25plus][LOCAL]   Done Refreshing
[DISP-#25plus][SERVER]   X!cservice@undernet.org has joined this channel
[DISP-#25plus][SERVER]   salt!~who@pylaser6.swan.ac.uk has joined this channel
[DISP-#25plus][SERVER]   dorothy!~dorothy@204.50.107.20 has joined this channel
[DISP-#25plus][SERVER]   OrbWeaver!~tom@darwin.cox.miami.edu has joined this channel
[DISP-#25plus][SERVER]   zmd!~jugarcia@dino.conicit.ve has joined this channel
[DISP-#25plus][washington.dc.us.undernet.org-MODE] Has changed #25plus's mode to +n
[DISP-#25plus][washington.dc.us.undernet.org-MODE] Has changed #25plus's mode to +t
[DISP-#25plus][washington.dc.us.undernet.org-MODE] Has changed X's mode to +o
[DISP-#25plus][washington.dc.us.undernet.org-MODE] Has changed *!*alexb@*.pcdocs.com's mode to +b
[DISP-#25plus][washington.dc.us.undernet.org-MODE] Has changed HOTGUY!scoto@ix-ftl2-17.ix.netcom.com's mode to +b
[DISP-#25plus][washington.dc.us.undernet.org-MODE] Has changed *!*John@*.sc.net's mode to +b
[DISP-#25plus][washington.dc.us.undernet.org-MODE] Has changed jwheeler!James@buffnet1.buffnet.net's mode to +b

[DISP-#25plus][washington.dc.us.undernet.org-MODE] Has changed Marcia_Br's mode to -o
[DISP-#25plus][SERVER]   electro-b!~Electro@ellen9.slip.yorku.ca has joined this channel
[DISP-#25plus][SERVER]   PaulaV!~pvcrazy@ppp162.iadfw.net has joined this channel
[DISP-#25plus]<Marcia_Br>   help!!!!
[DISP-#25plus][electro-b]   hello
[DISP-#25plus][salt]   hey hey marcia...back from the other side!!
[DISP-#25plus]<Marcia_Br>   what happened???
[DISP-#25plus][salt]   what's the matter marcia?
[DISP-#25plus][salt]   hi there electro
[DISP-#25plus][SERVER]   electro-b has left this channel
[DISP-#25plus][dorothy]   hey electro-b
[DISP-#25plus][dorothy]   hey marcia
[DISP-#25plus][OrbWeaver]   help .. marcia .. you got split.
[DISP-#25plus]<Marcia_Br>   i was lagged, logged off came back on and i was the only one here.
[DISP-#25plus][salt]   it was a split marcia...and you were on the "wrong" side
[DISP-#25plus]<Marcia_Br>   then all of this ban stuff came up.
[DISP-#25plus][zmd]   i see.. salt, i was just looking for a friend..
[DISP-#25plus]<Marcia_Br>   where was I?
[DISP-#25plus][OrbWeaver]   bad Marcia .. you were in the Twilight Zone.
[DISP-#25plus][salt]   on the chicago server hop

In this instance, had channel member, Marcia_Br, been on EFNet instead of Undernet, she would have had the potential to take over the channel as the split subsided. However, Undernet programmers designed a code system which specifically prevents this action, thus allowing for an orderly return of the channel to its 'normal' state, in which everyone knows the relative status of everyone else--there are no 'fake' ops.

A second important technical aspect of maintaining order is the creation of the "X" bot. What can occur without a channel bot? As discussed in Chapter Three, there are norms against certain behaviors, such as cursing, flooding and other generally inappropriate behavior. In a channel with no bot, the channel ceases to exist when everyone leaves. And the first person who re-enters the

Understanding Community Structure: Social Institutions and Groups

channel automatically becomes an op. Further, if there is an op in the channel who leaves without opping another trusted individual, then there is no recourse if those individuals are assaulted with improper behavior. In the following example, the channel #lesbian is a bot-less channel. It is a channel intended for women only, but here a male, Antipas, enters and floods the channel. The women in the channel have no recourse, although Nikole does attempt a counter-flood. She is eventually booted from the server for excessive flooding, and Antipas leaves the channel:

[DISP-#lesbian][SERVER] Antipas!TWWacc.wu@obelix.wu-wien.ac.at has joined this channel
[DISP-#lesbian][BAMBEE] Antipas,
[DISP-#lesbian][Antipas] For even their women did change the natural use into that which is against nature.
[DISP-#lesbian][Antipas] You pudendum eating she-dogs really should stop it.
[DISP-#lesbian][Antipas] You're headed for hell.
[DISP-#lesbian][BAMBEE] get lost anti
[DISP-#lesbian][Antipas] All that cunnilingus is not only bad for your breath, but it also is an assurance that you will one day wake up in hell.
[DISP-#lesbian][Antipas] For even their women did change the natural use into that which is against nature.
[DISP-#lesbian][Antipas] Who, knowing the judgment of God, that they which commit such things are worthy of death, not only do the same, but have pleasure in them that do them.
[DISP-#lesbian][Antipas] You dummies are worthy of death.
[DISP-#lesbian][Antipas] The only thing that could possibly save you is penitence.
[DISP-#lesbian][try]   ant-GET LOST!
[DISP-#lesbian][Antipas] And the only way that you can repent is with God.
[DISP-#lesbian][BAMBEE] go home anti
[DISP-#lesbian][Antipas] God has given you up to uncleanness,
[DISP-#lesbian][Antipas] vile affections,
[DISP-#lesbian][Antipas] and a reprobate mind.
[DISP-#lesbian][Antipas] That's why you're so perverse.

[DISP-#lesbian][BAMBEE] anti: people can do as they wish...God doesn't control what we can and cannot do!!!
[DISP-#lesbian][Antipas] That's why you think it's normal to do those things which are against nature.
[DISP-#lesbian][Antipas] You're nature freaks.
[DISP-#lesbian][Antipas] Hedonism is an abomination in the eyes of the Lord.
[DISP-#lesbian][Terri1] Anti And U are a weak-minded Zealot...GO Away!!!!
[DISP-#lesbian]<Marcia_Br> ignore it and it will go away...
[DISP-#lesbian][Antipas] You perverts think that if it feels good, then by all means do it.

[DISP-#lesbian][SERVER] nikole!nikole@onyx.southwind.net has joined this channel
[DISP-#lesbian][Antipas] But that isn't what the Bible teaches.
[DISP-#lesbian][Antipas] Deny all ungodliness and worldly lusts, dummies!!
[DISP-#lesbian][BAMBEE] Bible scmible...
[DISP-#lesbian][BAMBEE] nikole,
[DISP-#lesbian][Terri1] The OPs left w/o appointing new ones
[DISP-#lesbian][nikole] all real lesbians go to lesbians and leave mr bible beater from topeka here
[DISP-#lesbian][nikole] all real lesbians go to lesbians and leave mr bible beater from topeka here
[DISP-#lesbian][nikole] all real lesbians go to lesbians and leave mr bible beater from topeka here
[DISP-#lesbian][nikole] all real lesbians go to lesbians and leave mr bible beater from topeka here
[DISP-#lesbian][Antipas] That means when you feel like sucking on another female's pudendum, then you shouldn't just go out and do it.
[DISP-#lesbian][nikole] all real lesbians go to lesbians and leave mr bible beater from topeka here
[DISP-#lesbian][nikole] all real lesbians go to lesbians and leave mr bible beater from topeka here
[DISP-#lesbian][nikole] all real lesbians go to lesbians and leave mr bible
[DISP-#lesbian][SERVER] nikole has quit IRC Excess Flood

[DISP-#lesbian][Antipas]   For even their women did change the natural use into that which is against nature.
[DISP-#lesbian][Antipas]   For even their women did change the natural use into that which is against nature.
[DISP-#lesbian][Antipas]   For even their women did change the natural use into that which is against nature.
[DISP-#lesbian][SERVER]   Antipas has left this channel
[DISP-#lesbian][SERVER]   nikole has left this channel

If the organizers of a channel choose, for whatever reason, not to register the channel in Undernet and receive the protection of the X bot, they may still create a bot of their own which may serve various functions for channel members. However, these bots can 'crash', leaving the channel unprotected. For example, channel #gaychat is normally served by a bot named PoPaToP, which will op regulars when they enter the channel. Here PoPaToP has crashed, thus making it impossible for the ops to kick and ban TeleCon. Eventually, the bot returns to the channel and enforces the kick/ban commands issued to it by the ops:

[DISP-#gaychat][SERVER]   TELE-CON!~ckulbaba@node07.logicnet.com has joined this channel
[DISP-#gaychat][Timon]   hi telecon
[DISP-#gaychat][Southboy]   Guess graffiti isn't working today-
[DISP-#gaychat][TELE-CON]   Hey fags.
[DISP-#gaychat][KansasBoy]   !zap tele-con
[DISP-#gaychat][SERVER]   Doormouse!ez042222@dino.ucdavis.edu has joined this channel
[DISP-#gaychat][TELE-CON]   Z you licked little kids?
[DISP-#gaychat][Timon]   !zap tele-con what an ass
[DISP-#gaychat][SERVER]   PoPaToP has quit IRC Ping timeout for PoPaToP[irc.eskimo.com]
[DISP-#gaychat][KansasBoy-MODE]   Has changed telecon!*@*'s mode to +b
[DISP-#gaychat][Southboy]   Good job, KB!
[DISP-#gaychat][Timon]   kansasboy...good job!!!!!
[DISP-#gaychat][TELE-CON]   HEY YOU FAGS DONT BE LOOKIN AT MY ASS!

[DISP-#gaychat][SERVER]   Steven has left this channel
[DISP-#gaychat][Will]   hey type /server irc.escape.com
[DISP-#gaychat][TELE-CON]   FUCKING QUEERS.
[DISP-#gaychat][Will]   whoops
[DISP-#gaychat][TELE-CON-CTCP]   PING 3770
[DISP-#gaychat][Will]   Ignore that
[DISP-#gaychat][Alejandro]   Tele-con you probably don't have an ass to look at so don't even flatter yourself
[DISP-#gaychat][KansasBoy-MODE]   Has changed tele-con!*@*'s mode to +b
[DISP-#gaychat][Timon]   tele-con: you closeted homosexual
[DISP-#gaychat][Will]   OKIE!!!!
[DISP-#gaychat][Alejandro]   *giggle*
[DISP-#gaychat][Southboy]   Ignore him and kick him, ASAP!
[DISP-#gaychat][KansasBoy]   telecon, why are you pinging me?
[DISP-#gaychat][Southboy]   Is he kickable?
[DISP-#gaychat][Timon]   tele-con wants it bad in the ass
[DISP-#gaychat][SERVER]   TELE-CON has left this channel
[DISP-#gaychat][Alejandro]   dork

[DISP-#gaychat][Timon]   what a closeted fag
[DISP-#gaychat][KansasBoy]   I'll be right back...
[DISP-#gaychat][KansasBoy]   why were the /mode commands not working?
[DISP-#gaychat][ACTION]   Alejandro yawns
[DISP-#gaychat][SERVER]   Steven!~essmith@rigel.infinet.com has joined this channel
[DISP-#gaychat][Southboy]   !write Does this work?
[DISP-#gaychat][RUdd]   hey all
[DISP-#gaychat][KansasBoy]   the !zap command won't work if they are an opped user, but the /ban should
[DISP-#gaychat][SERVER]   KansasBoy has quit IRC
[DISP-#gaychat][Alejandro]   laterz all....I'm heading out
[DISP-#gaychat][SERVER]   Alejandro has quit IRC Leaving
[DISP-#gaychat][Southboy]   What happened to graffiti today?
[DISP-#gaychat][Steven]   ;dcc close chat popatop
[DISP-#gaychat][OkieBoy]   Popatop is gone.
[DISP-#gaychat][Steven]   Man, these freaking bots. When you need them.

[DISP-#gaychat][SERVER]   PoPaToP!ragtop@irc.eskimo.com has joined this channel
[DISP-#gaychat][PoPaToP] KansasBoy tries to invoke the powers of ^ ZAP^
[DISP-#gaychat][ACTION]   PoPaToP throws some powder to the ground and disappears.
[DISP-#gaychat][PoPaToP] ^ Poof^
[DISP-#gaychat][PoPaToP] Suddenly, tele-con's body rips from his legs and land 2 feet away.
[DISP-#gaychat][PoPaToP] You cannot kill what you cannot see.
[DISP-#gaychat][KansasBoy]   Yea... And every time that I try to op Steven it says that he is not in here...
[DISP-#gaychat][PoPaToP] Timon tries to invoke the powers of ^ ZAP^
[DISP-#gaychat][Will-MODE] Has changed PoPaToP's mode to +o
[DISP-#gaychat][ACTION]   PoPaToP pulls off his mask revealing a skull.
[DISP-#gaychat][ACTION]   PoPaToP looks at tele-con.
[DISP-#gaychat][PoPaToP] tele-con screams in pain as he is bathed in fire.
[DISP-#gaychat][PoPaToP] ^ *KABOOM*^
[DISP-#gaychat][PoPaToP-MODE] Has changed tele-con!*@*'s mode to -b
[DISP-#gaychat][PoPaToP-MODE] Has changed telecon!*@*'s mode to -b
[DISP-#gaychat][PoPaToP] ^ *ToASTy!*^
[DISP-#gaychat][Warlok]   hehehe

Institutions such as the bot (X in particular) are problem-solving strategies that allow for the maintenance of both community order and permanent individual identities in the absence of face-to-face visual cues. Whether or not this strategy is used in a given channel depends upon the experiences and values of those who form the channels. What follows are two explanations from the founders of two Undernet channels as to how the channels were started and the role of the bot institution. In the first example, the X bot is utilized. In the second example, the channel is bot-less, and order is maintained through a different strategy.
Source 1:

"To fully understand the reason why I started channel #anonymous, I will tell you a little about my encounters of people on IRC in general, and a little how I felt being a newbie. Because I think your later view and behavior on IRC is molded from your very first times on IRC and the people you meet.

It's a wonder how fast you get to know people on IRC, and as a newbie (at least in my case), I was a little shy so I didn't speak much in public on the channels. Instead I messaged people who I figured were nice judging from their public conversation. And after some weeks I had a net of friends on IRC, and most of us were all newbies more or less. So we stuck together and talked to each other about what we had experienced and what channels were good. We were a little social group in the whole IRC society.

After a couple of weeks on #family I felt good about being there. You felt like you were wanted in a way, and that feeling didn't come from the ordinary channel chat, but from the private messaging. That's the only way to get to know people and to make bonds which you feel secure about. The atmosphere in #family was friendly as you may understand and the group of regulars were pretty stable.

With the reminiscence of channel #family, and my personal attitude to people in general I had decided to start my own channel and make this channel the friendliest channel on IRC! To make a channel grow from 1 person (me) to a group of people who comes back so that would make the beginning of regulars, it's not enough to just be nice to everyone and it's not enough to have a good channel name that will attract people in a common fashion. I came up with the name #anonymous since there were already other age-related channels, and a channel name like that would hopefully tell people something about this channel. I would meet people of my own age and easier find common interests in life.

So what more will be needed to make a channel grow? Time! You must stay on your channel hours and hours everyday, in order to eventually get a little group of people who will come back. There is something magic about the number of people on a channel. If you are less than 5, then it wouldn't take long until new people will leave, because they feel it's a drag. Between 5-10 people and you can smile, because that is enough to keep a conversation on a channel and no silence, which will make new people leave in a jiffy. When you are more than 10 people, or should I say over 15 people something strange happens. Because the 5 biggest channels on IRC have a population about 20 people or more, you will find that more and more people join at a faster rate.

In the early days there was no bot on #anonymous, but I always had on mind that I must get a bot in order to protect my channel from jerks ruining the friendly channel or from people trying to take over my channel. This is one of the good reason to be an owner of a bot. After about 2 weeks and at least still spending about 5 hours everyday on #anonymous, I had collected my 10 supporters which is needed to register a channel.

A life was beginning to form on #anonymous. From nothing I had created a small society of friends all with one thing in common: there were all my friends, and everyone knew about me....and I had found a core of friends...and I had my X bot established on my channel. So I would say the most exciting things have now occurred on the channel...The people that at that time were my best friends are still operators everyone, and level 100. Which means that no one can add another operator to the permanent channel operator list, except from me. You will need access level 400. So there are about 10 people which I consider being my best friend on #anonymous." [Anonymous private email 1995]

In this description of the formation of a channel, it is clear how past experiences and values influence an individual's willingness to abide by community standards. Because of socialization experiences on #family, the leader of #anonymous valued the norms and belief system of the IRC community and chose to register #anonymous channel and gain access to the X bot in order to perpetuate these values and beliefs. Alternatively, channel members may choose not to utilize the X bot, socializing newcomers to IRC values through other means.

Source 2:

[DISP-#heat][Heatgain]   ok...this channel...on the undernet...was started by about 4 or 5 or so of us......because the effnet was getting too crowded....b4 that...it was started on the effnet..
[DISP-#heat]<Marcia_Br> how many were on effnet when you started undernet channel?

[DISP-#heat][Heatgain]   i'd guess around 50 or so....i have a list now of 112....and it's not a full list..
[DISP-#heat]<Marcia_Br> you mean there are 112 on undernet now?
[DISP-#heat]<Marcia_Br> or effnet?
[DISP-#heat][Heatgain]   well....112 regulars that gave me their rl names, addresses, phones....there's more than that on the channel, though..
[DISP-#heat][Heatgain]   on undernet.....i don't go to effnet anymore..
[DISP-#heat]<Marcia_Br> Wow! so you have a list of regulars? what do you do with it?
[DISP-#heat][Heatgain]   nothing, really....kind of like for invites to parties....emergencies...they give me the info knowing i wouldn't give it out unless they ok'd it..
[DISP-#heat]<Marcia_Br> how come you don't go to effnet anymore?
[DISP-#heat][Heatgain]   effnet seems to be full of.....er....rookies......and college kids....very annoying....plus the splits are horrendous..
[DISP-#heat]<Marcia_Br> oh. i've noticed it's a lot slower.
[DISP-#heat][Heatgain]   marcia...the effnet is overloaded...lags and splits all the time....plus....when people first get on the net...they get on effnet....toooo crowded..
[DISP-#heat]<Marcia_Br> oh, so undernet is more secluded and for more serious and experienced IRCers...
[DISP-#heat][Heatgain]   and the college kids....enjoy kicking and banning.....
[DISP-#heat]<Marcia_Br> hehe
[DISP-#heat][Heatgain]   and soon...we'll move on to another net..
[DISP-#heat]<Marcia_Br> really?
[DISP-#heat]<Marcia_Br> undernet too crowded, too?
[DISP-#heat][Heatgain]   yes...getting to crowded here too....there are already splinter channels..
[DISP-#heat][Heatgain]   too many effnetters finding their way down here..
[DISP-#heat]<Marcia_Br> i've noticed that even on undernet, some of the channels are immature
[DISP-#heat][Heatgain]   marcia...most channels are very immature...that's why people love ours..
[DISP-#heat][Heatgain]   altho ours is too, sometimes!
[DISP-#heat][Heatgain]   the other ones have...<especially with the college age kids>....kicking and banning going on all the time..

Understanding Community Structure: Social Institutions and Groups

[DISP-#heat]<Marcia_Br> yeah. so i take it there's not much of that on 41+?
[DISP-#heat][Heatgain] we only kick someone who insults or annoys any of our regulars..
[DISP-#heat]<Marcia_Br> but you still have the other norms that appear to be general IRC norms, like no cursing, flooding, stuff like that?
[DISP-#heat][Heatgain] well....cursing is tolerated if it's funny at that particular time....flooding and flashing are not tolerated..
[DISP-#heat]<Marcia_Br> what's flashing?
[DISP-#heat][Heatgain] flashing turns your screen into unreadable characters...you'd have to sign off and sign back on to clear it..
[DISP-#heat]<Marcia_Br> that sounds horrible!!!
[DISP-#heat][Heatgain] that's the principle reason we op one another...protection..
[DISP-#heat]<Marcia_Br> i've noticed that 41+ has LOTS of ops compared to other channels. why is that?
[DISP-#heat]<Marcia_Br> just for flashing?
[DISP-#heat][Heatgain] protection...kids....and morons....try to get ops....kick everyone off...or de-op everyone...and sometimes ban them or close the channel..
[DISP-#heat][Heatgain] lemme show you what used to happen...before we.....circled the wagons..
[DISP-#heat]<Marcia_Br> ok
[DISP-#heat][Heatgain-MODE] Has changed Marcia_Br's mode to +o
[DISP-#heat][Heatgain] ok...you're an op...right?
[DISP-#heat]<Marcia_Br> right.
[DISP-#heat][Heatgain] ok...let's say you started this channel...and deserve to be an op..
[DISP-#heat]<Marcia_Br> o.k.
[DISP-#heat][Heatgain] this used to happen...
[DISP-#heat][Heatgain-MODE] Has changed Marcia_Br's mode to -o
[DISP-#heat][Heatgain-MODE] Has changed *!*m@*.phxslip4.indirect.com's mode to +b
[DISP-#heat][LOCAL] Heatgain!heatgain@sefl.satelnet.org has BANNED US by *!*m@*.phxslip4.indirect.com
[DISP-#heat][Heatgain-KICK] has kicked Marcia_Br from #heat ^ l PhoEniX l^
[DISP-#heat][Heatgain] see?

[DISP-#heat]<Marcia_Br> yeah. that would suck. did that happen a lot on 41+?
[DISP-#heat][Heatgain]   i could also change the mode of the channel...
[DISP-#heat][Heatgain-MODE]   Has changed #heat's mode to +i
[DISP-#heat][Heatgain]   see?
[DISP-#heat]<Marcia_Br> yep.
[DISP-#heat][Heatgain-KICK] has kicked Marcia_Br from #heat ^ l PhoEniX l^
[DISP-#heat][Heatgain]   see?
[DISP-#heat]<Marcia_Br> yeah.
[DISP-#heat]<Marcia_Br> you need to be really careful.
[DISP-#heat][Heatgain]   and you aren't banned...i just made it by invite only..
[DISP-#heat][Heatgain-MODE]   Has changed #heat's mode to -i
[DISP-#heat]<Marcia_Br> oh, o.k.
[DISP-#heat][Heatgain]   this would happen a lot...specially on the effnet..
[DISP-#heat][Heatgain]   that's how the system of opping all regulars came about..
[DISP-#heat]<Marcia_Br> gotcha.
[DISP-#heat][Heatgain]   yes...if i saw you start to de-op people....i would try and deop you before you deopped everyone..
[DISP-#heat]<Marcia_Br> has anyone tried to do that recently? take over the channel i mean.
[DISP-#heat][Heatgain]   yes.....a recent drop in from the effnet....they ARE ways to get ops without an op opping you..
[DISP-#heat]<Marcia_Br> really? how do you do that?? i thought you could only get it from another op or a bot.
[DISP-#heat][Heatgain]   well...you can ride in on a netsplit..
[DISP-#heat]<Marcia_Br> hehe
[DISP-#heat]<Marcia_Br> yes, this all makes good sense now...41+ having a lot of ops I mean.
[DISP-#heat][Heatgain]   yes....covering each others butts ...and protecting our channel..
[DISP-#heat]<Marcia_Br> how come no bot? not allowed?
[DISP-#heat][Heatgain]   no...we don't like bots.....
[DISP-#heat]<Marcia_Br> why not?

[DISP-#heat][Heatgain]   there are other channels we registered....they have bots...are we are registered with the bots...that's where the 50 to 500 come in..
[DISP-#heat]<Marcia_Br> o.k.

Because channel #41plus on the Undernet was a channel which splintered from its EFNet counterpart, channel leaders felt that the best way to maintain community values was to develop a list of regulars who could be trusted with ops. Since it is a very crowded channel, one can always trust that it will never be op-less as was #lesbian in the example above. Channel leaders chose not to register and receive the X bot because they wanted all ops on equal footing, rather than on various levels of authority. Regardless, the same basic goals are achieved.

Besides the bot, IRC channel members have developed a number of other strategies for maintaining social order. Two such strategies are the naming of channels and the use of private channels. Selection of a good name is critical to a channel's success because, in a text-based reality, such descriptors become key elements that participants use in defining a group identity.

The channel name allows an outsider to understand what the discussion will be about and what types of people will be involved before he or she ever enters the channel. It is instrumental in selecting which channels to enter and which ones to stay away from, as it allows one to impute a whole range of personal characteristics and motives to the individuals involved. Channels are usually named by age groupings, ethnic or sexual identity, or by topic (for example #netsex, #amiga or #startrek). Without such categorizations, it would be impossible for any order to be established in Undernet. Likewise, it would be difficult for individuals to develop a sense of identity or community with others as the channel could potentially remain unstructured. Thus channel names are the strategies through which identities and situations are arranged and maintained.

Another important institutional strategy is the private channel, which is instrumental in developing more intimate bonds among individual channel members. Because the general atmosphere of the Undernet is that of the Third Place, it is generally not considered appropriate to delve into lengthy discussions of a personal nature in the public channels. However, individuals can and do utilize private channels for this reason, and it is often the case that one encounters, and reciprocates with, a very high degree of self-disclosure in these private situations as a means for establishing a sense of the other's 'true self', as the following example illustrates:

[p] want to hear my first experience online??
<M>   ok
[p] before I came here I was on another system, a national service
[p] I was a sysop
[p] which means I managed a particular area
<M>   ok
[p] anyway, I met a woman from NY
[p] we were both 'working' in the same area
[p] she was (is) married
[p] has 2 kids
[p] well, when I met her she was really down and sad
[p] and in a lot of pain from a broken online relationship with a guy
[p] and she cried and cried all the time
[p] we became good friends as time passed
[p] I am something of a "caretaker" ...
[p] and tried to be there for her a lot, if you know what I mean
<M>   yeah
[p] to be a friend and comfort her
[p] as time passed .....
[p] the friendship deepened ....
[p] and I told her of my sexuality
[p] (I'm much more open now than I was then)
[p] well, we just seemed to get closer
[p] and she tells me of a childhood experience she had ..
[p] of touching the breasts of her friend.
[p] Not unusual at all ... most kids explore some
[p] she goes on and on ...
[p] of how she would like to do this again :)
[p] so ......
[p] we talk a lot about this, and we grow closer because we are just comfortable, etc.
[p] I became very attached to her
[p] and I began to care very deeply for her
[p] But ... she couldn't quite feel for me what I thought I felt for her

[p] so I was pretty hurt about it all
[p] so .. this is one example of someone with bi feelings
[p] but she really is hetero
<M>   I'm sorry.
[p] if it is possible to fall in love online, I did
[p] and I still think about her very often
<M>   what happened to her?
[p] she's still married, and having online flings with men I suppose
[p] I was very hurt, and ended the relationship
<M>   oh. i'm really very sorry......
[p] thanks :)
[p] I'm OK with it now ...
<M>   was this a long time ago?
[p] but I will be careful for now on :))
<M>   i don't blame you:)
[p] year and a half?
[p] before that, I had another online relationship
<M>   that's not that long ago.
[p] someone I met from Tx on Prodigy
<M>   did it have a better outcome, I hope?
[p] and we met twice
[p] but it didn't last
<M>   oh:(
[p] and since then, I've had some attachments, but none that were of the painful nature!
<M>   that's good:)
[p] I get propositioned often, but just not interested right now
<M>   I don't blame you:)

Additionally, private channels are used by individuals to evaluate situations and/or individuals in the public channel that they currently occupy. This is often done in the case of a problematic, or deviant, situation or individual. The situation or individual is assessed by others in a behind-the-scenes manner, and depending on who is doing the assessing, such private channel conversations may impact the public channel (e.g., if two channel ops are assessing an individual in a private channel for purposes of deciding whether to kick/ban him or her). In any event, such channel conversations serve to establish order in public channel situations, as community members work together to evaluate a problem and choose an appropriate solution.

Another way in which community members maintain social order is by using a system of nicknames, or nicks. One who is inexperienced at IRC might consider it a 'stretch' to suggest that the nick is an institution. But to the people involved in IRC, the nick is the key means through which identity is established; members use the nick to assess a newbie's personality in much the same way as people in face-to-face interactions use personal appearance to generate a 'first impression'. While some people use their 'real names' as nicks, most choose a nick that they feel describes something essential about their personalities. Because the nick may only utilize nine characters, individuals must be creative in getting their statements of personality across to others. This sense of the nick representing the entire person is most easily understood through the use of an example of poor nick choice:

[DISP-#35plus][SERVER]   REALHUNK!~jlaksdl@204.7.222.140 has joined this channel
[DISP-#35plus][LComeno] bad nick
[DISP-#35plus]<Marcia_Br>   lol
[DISP-#35plus][Elinore]   heheee!!
[DISP-#35plus][SERVER]   REALHUNK has left this channel
[DISP-#35plus]<Marcia_Br>   gee, and just when i was going to go for it. hehehe
[DISP-#35plus][ProTalk]   Ladies you missed REALHUNK ...
[DISP-#35plus][LComeno] why do they think a nick like that will help them
[DISP-#35plus][ACTION]   Elinore sooooo upset...missed realhunk
[DISP-#35plus][jeff2]   that used to be my name...
[DISP-#35plus]<Marcia_Br>   lcomeno: who knows....
[DISP-#35plus][ACTION]   LComeno was getting ready to kick realhunk

In this instance, 'realhunk' is immediately singled out as a potential threat to the group's values based solely on poor nick selection. Not only did both males and females in the channel ridicule the name, but Lcomeno, a channel op, was prepared to kick 'realhunk' based solely on his nick.

In fact, channel operators (ops) have a number of specialized 'powers' available to them that they use to maintain social order. Included in these powers are the ability to elevate other members to 'op' status, make a channel private, and even to kick out and permanently ban individuals from their channels, as the following example illustrates:

DISP-moth][moth]   do you know how to be an op?
[DISP-moth]<Marcia_Br>   no. i wish i did. but i'm pretty new.
[DISP-moth][moth]   do you have to go right now? i could teach you
[DISP-moth]<Marcia_Br>   well, o.k. is it hard?
[DISP-moth][moth]   no, i'll just give you the basics
[DISP-moth]<Marcia_Br>   o.k.
[DISP-moth][moth]   first join #moth so we don't have to do this all in msgs :)
[DISP-moth]<Marcia_Br>   o.k. just join the channel?
[DISP-moth][moth]   yeah, type /join #moth
[DISP-moth]<Marcia_Br>   o.k. wait a sec...

[DISP-#moth][moth]   first, to make someone an op (once you are one) you do /mode #channelname +o nick
[DISP-#moth][moth]   so to op you i'll type /mode #moth +o marcia_br
[DISP-#moth][moth-MODE]   Has changed Marcia_Br's mode to +o
[DISP-#moth][moth]   see that?
[DISP-#moth]<Marcia_Br> yeah!
[DISP-#moth][moth]   okay, now i want you to de-op and then op me, by doing first -o and then +o
[DISP-#moth]<Marcia_Br> o.k.
[DISP-#moth][Marcia_Br-COMMAND]   mode #moth -o moth
[DISP-#moth][Marcia_Br-MODE]   Has changed moth's mode to -o
[DISP-#moth][moth]   excellent!
[DISP-#moth]<Marcia_Br> o.k. now i'll try to change you back.
[DISP-#moth][Marcia_Br-COMMAND]   mode #moth +o moth
[DISP-#moth][Marcia_Br-MODE]   Has changed moth's mode to +o
[DISP-#moth]<Marcia_Br> cool.
[DISP-#moth][moth]   okay, now for other modes
[DISP-#moth][moth]   to make the channel invite only, you type /mode #channelname +i
[DISP-#moth][moth]   to make it private, it's +p
[DISP-#moth]<Marcia_Br> o.k., but then how do you invite people?
[DISP-#moth][moth]   by typing /invite nick

[DISP-#moth]<Marcia_Br> what's the difference between invite only and private?
[DISP-#moth][moth]   private just means that it doesn't appear when someone not on-channel does /list
[DISP-#moth]<Marcia_Br> ohhh. when would you want to make it private?
[DISP-#moth][moth]   and when someone does a /whois on you (and is not on-channel) they won't see the name of that channel
[DISP-#moth]<Marcia_Br> oohhh. o.k.
[DISP-#moth][moth]   if you only want people who already know about the channel to come
[DISP-#moth]<Marcia_Br> o.k.
[DISP-#moth][moth]   #lesbians is private
[DISP-#moth][moth]   you should check out that channel too, it's a great one
[DISP-#moth]<Marcia_Br> oh. then how did i find it? i've never been
[DISP-#moth]<Marcia_Br> oh, was i on lesbian, not lesbians?
[DISP-#moth][moth]   no, #lesbianS not #lesbian
[DISP-#moth]<Marcia_Br> o.k.
[DISP-#moth][moth]   yeah, exactly
[DISP-#moth][moth]   okay, got it all so far?
[DISP-#moth]<Marcia_Br> i think so.
[DISP-#moth][moth]   to change the topic on a channel you just do /topic whatever you want the topic to be
[DISP-#moth]<Marcia_Br> o.k.
[DISP-#moth][moth]   so try /topic the learning channel
[DISP-#moth][Marcia_Br-COMMAND]   topic #moth the learning channel
[DISP-#moth][SERVER]   Marcia_Br has changed topic to the learning channel
[DISP-#moth]<Marcia_Br> ooohhhh.
[DISP-#moth][moth]   see? easy and fun :)
[DISP-#moth]<Marcia_Br> yeah:)
[DISP-#moth][moth]   to make it so that only the op can change the topic, make the channel +t
[DISP-#moth]<Marcia_Br> what does 't' stand for? topic?
[DISP-#moth][moth]   okay, now for kicking and banning, the best part
[DISP-#moth][moth]   yeas
[DISP-#moth]<Marcia_Br> hehehe
[DISP-#moth][moth]   yes, i mean

Understanding Community Structure: Social Institutions and Groups 69

[DISP-#moth][moth]　　to kick someone, all you have to do is /kick #channelname nick
[DISP-#moth]<Marcia_Br> if there are many ops on a channel, who decides to do what?
[DISP-#moth][moth]　　so try kicking me
[DISP-#moth][moth]　　any op can do any of this
[DISP-#moth][moth]　　so, depends on what you want to do
[DISP-#moth]<Marcia_Br> well, shouldn't you tell me how to get you back first??
[DISP-#moth][moth]　　i can get back just by rejoining
[DISP-#moth]<Marcia_Br> oh, they don't have to all agree?
[DISP-#moth]<Marcia_Br> o.k. here goes.
[DISP-#moth][moth]　　no, so don't give out ops unless you trust someone
[DISP-#moth][Marcia_Br-COMMAND]　　Kicked moth from #moth
[DISP-#moth][Marcia_Br-KICK]　　has kicked moth from #moth Marcia_Br
[DISP-#moth][SERVER]　　moth!FIG@bos1g.delphi.com has joined this channel
[DISP-#moth]<Marcia_Br> do you get any message when you're kicked?
[DISP-#moth][moth]　　and if you want to add a message to send them on their way, you just type that after the nick
[DISP-#moth]<Marcia_Br> o.k. hehehe
[DISP-#moth][moth]　　like /kick #moth moth see ya!
[DISP-#moth][Marcia_Br-COMMAND]　　mode #moth +o moth
[DISP-#moth][moth]　　now, is someone is a real problem and you don't want them to come back at all, you can ban them
[DISP-#moth][Marcia_Br-MODE]　　Has changed moth's mode to +o
[DISP-#moth]<Marcia_Br> how's that?
[DISP-#moth][moth]　　thanx!!
[DISP-#moth]<Marcia_Br> do you just replace 'kick' with ban and the rest is the same?
[DISP-#moth][moth]　　to ban you i'd type /mode #moth +b m@s152.phxslip4.indirect.com
[DISP-#moth][moth]　　you have to use the full user@domain
[DISP-#moth][moth]　　and to remove the ban, it is the same thing only -b
[DISP-#moth][moth]　　sometimes you get major jerks who won't leave the channel alone
[DISP-#moth][moth]　　and by banning them, they are out of there for good

[DISP-#moth][moth]	but you shouldn't have to use it much
[DISP-#moth][moth]	so, there you go
[DISP-#moth][moth]	that's basically it

Though channel ops have certain specialized powers, all channel members have available to them a wide range of system commands that they may use to further personalize interactions or to verify another user's identity. These commands are preceded by the "/", and this fact is utilized, along with the names of some commands, to reference the peculiarities of IRC life. For example, the "/ping" command is used if one suspects that she or he is suffering from lag. An individual can /ping another individual or an entire channel and learn the distance (in the number of seconds) that she or he is 'apart' from the other. So, if one "/ping user1" and the response is "pong 28" seconds, this means that one is 28 seconds away from user1 and it will take that long to see what they type on the screen. This is a fairly severe lag. Brief games of /pinging others frequently occur in which /ping is referred to in a sexual sense; jokes are made about the quality and frequency of /pings.

The "/message" command is used when one wishes to send a message through a private channel. One would issue the command "/message, nick, text of message". Undernet participants also make heavy use of the "/action" command. This allows participants to simulate physical activities and facial expressions in order to create atmosphere for the channel. In addition, the /action command is also used to make statements and ask questions in a manner which is acceptable to the culture of the Third Place.

Finally, there are several commands that are strategies designed to help establish and maintain personal identity in IRC. These are "/nick","/whois", "/whowas", and "/finger". The /nick command allows an individual to change his or her nickname. It is commonly assumed because of this that members of the Undernet community have no means of knowing who someone 'really' is. But a change in the nick does not erase all of the information relevant to that IRC identity. An op can kick a nick, but the nick can return immediately to the channel. In order to permanently ban an individual, one must know the site information. Because the ban is based on the more permanent identity of the site one is using, simply changing one's nick will not allow the individual back into the channel from which she or he has been banned.

Such site information is gained through the /whois, /whowas and /finger commands. If any channel member wants information on a nick, he or she can type "/whois nick" and the site the user is on and often times the person's real name are revealed. The /whowas command supplies the same information, but is designed to be used immediately after the individual leaves the channel. The /finger command is similar to these. However, it can potentially supply the inquiring individual with more information than the /whois command does. Additionally, when the /finger command is utilized, the person being /fingered knows it. This is in contrast to the /whois command, which one can do without the subject of the /whois realizing it. In fact, this use of the /finger command can in and of itself be used as a means of social control, because it informs the 'deviant' that he or she is being closely scrutinized.

Though this is merely an overview of the myriad strategies that IRC community members use to organize and maintain successful channels, one does get the sense that these groups, though relatively young, are very evolved in their ability to meet basic human community needs.

THE USENET COMMUNITY

The same argument may be made for the Usenet community. Over time, participants have collectively devised several strategies for dealing with community-wide problems. But, unlike the IRC community, Usenet's institutional strategies are more elaborate and well-documented, and tend to have more influence over the day-to-day newsgroup operations than do the community-wide processes of the Undernet.

Perhaps the most widely used strategy is the FAQ, or frequenty-asked-questions forum. Unlike the Undernet, where the purpose of interaction is to 'hang out', Usenet newsgroups are designed to allow users to discuss specific issues of interest to them; hanging out and asking 'how to' questions or questions regarding life on Usenet is considered a waste of time and bandwidth that violates the cultural emphasis on efficiency. Thus, the problem is how to socialize Usenet newbies to the rules without 'wasting time' within newsgroup interaction on answering such questions.

The creation of FAQs solves this Usenet-wide problem. FAQs have been created for every conceivable cluster of issues that require frequent explanation, and, because these FAQs are referred to with frequency by Usenetters, it is relatively easy for a newbie to locate the relevant FAQs. According to the news.announce.newusers newsgroup and the Usenet Help

World Wide Web page, there are six FAQs that are generally considered to be a mandatory course of 'required readings' for Usenet newbies: 1) A Primer on How to Work With the Usenet Community; 2) Answers to Frequently Asked Questions About Usenet; 3) Emily Postnews Answers Your Questions on Netiquette; 4) Hints on Writing Style for Usenet; 5) Rules for Posting to Usenet; and 6) What is Usenet? The Primer on How to Work with the Usenet Community lists 19 rules of Usenet netiquette for the newbie to follow, and stresses that these rules are designed to uphold the Usenet values of efficiency and politeness:

"This message describes the Usenet culture and customs that have developed over time. . .This document is not intended to teach you how to use Usenet. Instead, it is a guide to using it politely, effectively and efficiently. Communication by computer is new to almost everybody, and there are certain aspects that can make it a frustrating experience until you get used to them. . The easiest way to learn how to use Usenet is to watch how others use it. Start reading the news and try to figure out what people are doing and why. . ." [Von Rospach 1994, 1]

The Answers to Frequently Asked Questions About Usenet document is a conglomeration of answers to several different types of questions and topics that ". . .occur repeatedly on Usenet. They frequently are submitted by new users, and result in many follow-ups, sometimes swamping groups for weeks. The purpose of this note is to head off these annoying events by answering some questions and warning about the inevitable consequence of asking others." The questions in this document range from technically oriented 'how-to' issues to questions about Usenet mythology.

The Emily Postnews FAQ regarding netiquette as mentioned in Chapter Three, is one in a series of Emily Postnews FAQs and is constructed in a satirical format in order to emphasize the importance of adhering to social norms designed to promote efficiency and politeness. The document discusses such issues as signature files; improper use of Usenet to post mail; test messages; postings regarding incorrect information; summarizing previous articles; selection of groups to post in; newsgroup creation; spelling; subject selection; tone of articles; and jokes, among other topics.

The Hints on Writing Style FAQ lists several 'pointers' on writing style and on on-line interaction style that all Usenet users should take into account when posting articles to Usenet newsgroups. Within the section on writing style, the author suggests the following: 1) write below the readers' reading level; 2) keep paragraphs short and sweet; 3) white space is not wasted space; 4) pick your words carefully; 5) people can only grasp about seven things at once; 6) avoid abbreviations and acronyms; 7) use active voice; and 8)'cute' misspellings are difficult to read.

Within the section on on-line interactions, the author points out: 1) subtlety is not communicated well in written form; 2) humor is not communicated well either; 3) take a break before posting something in anger; 4) subject lines should be used carefully; 5) references need to be made; 6) do not include the entire article you are replying to; 7) it's easier to read a mixture of upper and lower case letters; 8) leaving out articles ('a' or 'the') confuses the reader; 9) be careful of the contextual meanings of words; 10) make an effort to spell correctly; 11) use subheadings to organize long articles; 12) re-read articles before posting; 13) remember Usenet is an international network; and 14) remember that people who are important to you may be reading your articles and will long-remember your gaffes.

Similarly, the Rules for Posting to Usenet FAQ is devoted strictly to netiquette regarding when and how one should post articles and to what newsgroups. The topics discussed include: 1) use of the 'distribution' feature to restrict one's article to appropriate newsgroups; 2) determining the expiration date of one's article to conserve bandwidth; 3) postings regarding major news events or professional products and services; 4) rules for posting to moderated newsgroups; 5) rules regarding copyrights and the posting of published works and private email; 6) postings of signature files; 7) canceling posted articles; 8) the appropriate circumstances for posting questions to newsgroups; 9) rules for cross-posting articles; 10) posting follow-up articles; and 11) appropriate article preparations.

Finally, the What is Usenet FAQ, is an attempt to define and explain Usenet to newbies. It offers a description of Usenet, a discussion of why it is difficult to define Usenet, and a discussion of what Usenet is not.

These FAQs are generally accepted throughout the Usenet community as the rules of proper net behavior to which each individual member should adhere. They are a primary means of socialization and social control within Usenet; "One of the ways to exert control over the workings of the net is to take the time to put together a relatively accurate set of answers to some frequently asked questions and post it every month. If you do this right, the article will be stored for months on sites around the world, and you'll be able to tell people, 'don't ask this question until you've read the FAQ.'" [What is Usenet, A Second Opinion 1994, 6]

Besides netiquette-oriented FAQs, Usenet members have also created FAQs to deal with the other key community-wide problem; how to create new newsgroups. Not only have Usenetters developed FAQs to inform individuals of newsgroup creation rules, they have also developed several important group processes in order to enforce those rules.

There are two main FAQs that deal with the creation of new Usenet newsgroups; the How to Create a New Usenet Newsgroup FAQ and the Usenet Newsgroup Creation Companion. According to the Usenet Newsgroup Creation Companion:

> ". . .there is a general agreement among Usenet news administrators that groups in the "big seven" (comp., misc., news., rec., sci., soc., talk.) hierarchies will only be honored at their sites if the group passes the "official" voting procedures defined in the Guidelines. Anyone can create a group if they figure out the correct message format to do so, but it will only be carried on a minuscule number of sites, and anyone posting to the group may be greeted with messages claiming that the group is bogus.
> But the Guidelines aren't the whole story. Another set of customs has sprung up around newsgroup votings - mostly because there has been an unfortunate number of sleazy tactics used in past group creation attempts. As with Caesar's wife, a vote must be beyond reproach. If you even accidentally violate one of these customs, you may find yourself with a botched vote, a lot of wasted time, and a massive flamewar. The purpose of this document is to help you through this potential mine field." [p. 1]

There are five main stages to this voting process: 1) Discussion of the idea to see if there is enough interest on the part of others to warrant the creation of a new group; 2) RFD (Request for Discussion), in which the group proponent posts an RFD to all interested groups plus news.announce.newgroups and news.groups (if net.announce.newgroups is not included in the RFD distribution, it is not considered an 'official' RFD); 3) Discussion in the news.groups.* hierarchy, during which time Usenet members respond to the RFD by expressing support, opposition or offering suggestions. (If the RFD is 'significantly altered' based upon such discussion, it is recommended that a second RFD be issued); 4) Voting, for which one contacts the Usenet Volunteer Votetakers Group within the pre-defined time limit; and 5) Results, in which the votetaker compiles the results and posts them to the net. If the group passes, it will be newgrouped after five days. If the group fails, it can't be voted on again for six months. [Usenet Newsgroup Creation Companion, pp. 1-3]

Usenetters have created a group known as the Usenet Volunteer Votetakers, which oversees the voting process in new newsgroup creation. According to the Usenet Volunteer Votetakers Information Center:

"The Usenet Volunteer Votetakers (UVV) group performs all the votetaking tasks for proposed Usenet newsgroups. The UVV was founded in August 1993 by votetakers who had been reliable and run successful neutral votes. UVV members believe in upholding the highest standards of ethics and professionalism in Usenet voting. The group has been designated as the sole votetaking group by the moderator of news.announce.newgroups and the members of the group-advice mail list."

Besides the UVV, the other community-wide group which works to maintain the stability and integrity of the Usenet newsgroup hierarchy is the Usenet Group Mentors. This group pairs a new group proponent with an individual who is experienced in the process of group creation. This mentor guides the new proponent through the process in order to ensure a fair outcome which will be accepted by site administrators, thus ensuring acceptance and propagation of the new group should the vote pass. Specifically, the Group Mentors are firm, diplomatic, responsible individuals skilled in the selection of appropriate group names that will fit within the Usenet namespace hierarchy, and the writing of appropriate RFDs. Below is a sample RFD for the creation of the humanities.language.comparative.african newsgroup:

REQUEST FOR DISCUSSION (RFD)
  moderated group humanities.language.comparative.african
This a formal Request For Discussion concerning the creation of a new moderated newsgroup called humanities.language.comparative.african. This is not a Call For Votes (CFV). Please do not vote at this time.
RATIONALE: humanities.language.comparative.african

Research on African Languages; all with the exception of Bantu, is at present non-existent in discussions carried out within Newsgroups or on mailing lists. This is also true of comparative research on African Languages.

It is for this reason that a group of comparative linguistics specialists dealing in African languages of the Sahel region (http://lolita.unice.fr/~rn/GDRE_orig.html) proposes to create a newsgroup entitled: humanities.language.comparative.african

The group formed as a network around a research project concerned with lexical distribution in the Sahel-Sahara zone, as well as with the study of linguistic connections among languages in the region. It is a French CNRS research organism.

CHARTER: humanities.language.comparative.african

As a moderated group, humanities.language.comparative.african would be open to topics where comparative linguistics in african languages are the subject of discussion, this would include :
- languages and language families present in the Sahel-Sahara zone: Mande, Chadic, Berber, Nilo-Saharan...
- Analysis of results obtained in the research on genetic relationships.
- Problems related to the description of changes in the context of languages of oral tradition.
- Understanding linguistic changes and all the relevant factors concerning language transformation.

Research workers who do not specialize in the above-mentioned languages, but who nevertheless are interested in general issues of theoretical or methodological value for the work carried out or research tools involved, are welcome to contribute.

Issues of the following type will thus be discussed:
- Empirical issues (Particular facts pertaining to a language or a group).
- Methodological issues (Procedures and methods of description).
- Theoretical issues (Principles of language change).
- Epistemological or cognitive issues (The validity of established facts and how these are obtained).

Contributions can be posted either in English or in French.

Moderation Policy: Contributions liable to be refused :
1) off-topic articles
2) gross profanity and indecency
3) binaries

4) advertising
5) non-specialist articles
END CHARTER.
MODERATOR INFO: humanities.language.comparative.african
humanities.language.comparative.african will be a moderated group. The group has designated Robert Nicolai, initiator and director of the project, as Moderator.
PROCEDURE:
This message initiates a discussion period to consider the creation of a humanities.language.comparative.african newsgroup. Discussion will take place on news.groups. If discussions are made in other newsgroups, they should always be cross-posted to news.groups.
* This is not a call for votes. Please do not attempt to vote now.
A call for votes (CFV) will be issued approximately 4 weeks after this
RFD. When the CFV is posted, there will be instructions on how to mail
your votes to the independent vote-taker.
DISTRIBUTION:
This RFD is in accordance with the Guidelines for Newsgroups Creation, and has been cross-posted to the following relevant newsgroups:
news.announce.newgroups,news.groups,sci.lang,soc.culture.nigeria, soc.culture.africa.

After this proposal has been discussed for the appropriate length of time, a call for votes is issued. All interested parties cast their votes, and wait for the results. Community members take the maintenance of newsgroup integrity very seriously. If one intends to start a new newsgroup, one must follow all procedures properly, or the newsgroup will not be considered legitimate and will not be propagated by site administrators. These rules for newsgroup creation and the groups that have developed to facilitate such activities have evolved over time to meet the needs of the community as it has grown more complex. In order to create a newsgroup, one must develop a Rationale and a newsgroup Charter that are capable of justifying the need for the new group. All groups which may be affected by the event must be put on the distribution list so that members may have ample time to discuss and debate the relevant issues. This Rationale and Charter must be explicitly stated at all stages in the voting process in order to ensure that all affected Usenetters have ample opportunity to fully understand the potential implications.

In general then, the FAQ system and the newsgroup creation-related processes that Usenetters have developed are community-wide attempts to uphold the values and social norms of Usenet, and members have developed these processes for the same reasons that members of face-to-face community members do; to maintain a sense of cultural identity and social order and to socialize new members to the values and behavioral norms of the group.

Because the technology of Usenet allows a single individual to potentially affect the entire community by cross-posting messages to every single newsgroup (unlike IRC, in which postings are limited to a single channel), Usenetters have a vested interest in documenting and making explicit the rules of netiquette and group creation. The potential for any single individual to disrupt the net, along with the cost structure of Usenet combine to create circumstances under which the potential for newsgroup disruption is very high. It is necessary under such circumstances for community members to develop the means by which they can protect the community, its values and ways of life from those who are potential threats. Usenetters are successful in their strategies; their community is marked by a remarkably high degree of order.

Additional community-wide strategies include the newsgroup 'Big Seven' hierarchy and Usenet namespace. As with Undernet channel names, the naming of newsgroups provides a social frame through which Usenetters may define the purpose of the group, and what specific activities constitute appropriate and inappropriate behaviors. As McLaughlin, Osborne and Smith [1995] note, "Virtual communities of interest can form only if everyone adheres to a common set of guidelines for organizing the millions of posted messages into a coherent set of groups, subgroups, and topics." [p. 105]

This 'common set of guidelines' is known as the 'Big Seven' hierarchy. Usenet was originally designed to link two university computer systems together. As the system continued to grow, users were 'forced' by the volume of postings to develop a means for organizing and distributing postings with greater efficiency. According to The Great Renaming FAQ [1995]:

> "By 1986, Usenet was experiencing some growing pains. The original scheme of just three worldwide hierarchies . . . was becoming difficult to administer. . . The Great Renaming started July 1986 and ended in March 1987. . . . Much of the debate centered on ways in which the wider Usenet community could somewhat support the

backbone financially, so everyone could keep getting the groups they wanted. . . In any event, after much debate and revision, the current comp.*, misc.*, news.*, rec.*, sci.*, soc.* and talk.* hierarchies were created.

Yet ironically, nothing really changed for a significant portion of sites, and almost all the sites except company ones still got a full feed. As has been so often the case in Usenet history, its "imminent death" was postponed by advances in technology. . .

The original system eventually became the proposal and 'vote' scheme of the Big 7. Holding votes was a way to make people shut up if a group was unpopular. Even though there is no one 'in control' of Usenet, the voting system is the closest thing to a government it has. The current vote format was laid down right before the Great Renaming, but not used until 1987. Once the voting mechanism was in place, the individual opinions of the backbone-site admins no longer mattered much. . ." [pp. 2-4]

So, in order to accommodate a growing population of users and posts, the original creators of Usenet chose to restructure not only the hierarchy of newsgroups themselves, but the means by which such newsgroups would be created in the future. The Big Seven hierarchy serves to facilitate message transfer by sysadmins. But, more importantly, it serves to establish legitimate boundaries of interaction across Usenet as a whole in two specific ways. First, the Usenet newsgroup hierarchy, and all corresponding newsgroup namespaces, allow participants in a particular newsgroup to define themselves as members. The development of a legitimate newsgroup categorization system suggests the legitimacy not only of the group, but of identification with that group as a cohesive unit.

Second, not only can one claim membership in a particular, recognized group, one can also establish what types of posts and interactions are acceptable in that group. If a newsgroup is contained within the rec.* hierarchy, for example, then it is understood by all Usenetters that posts transmitted to that hierarchy should be relevant to the mission of that hierarchy--arts, hobbies and recreational activities. Further, if the newsgroup is contained within rec.travel.*, then messages posted to it should be relevant to recreational traveling, and so forth. It is because Usenetters utilize the Big Seven hierarchy and namespace conventions that members can define certain postings as 'appropriate' and others as 'spam'.[2] They can define the tone of articles as acceptable or unacceptable; location in the hierarchy determines which postings are flames and which are not, because what should be considered a flame is highly contextual.

As Kollock and Smith [1994] note, "In many ways, a newsgroup's name is one of its most effective means of defining a boundary: by announcing its contents it attracts the interested and repels the disinterested. But within this boundary a newsgroup's membership can be extremely fluid. Some newsgroups do attract and hold a fairly stable group, but many do not. To the extent membership in a newsgroup isn't stable and its boundaries are not clearly defined, cooperation will be more difficult." [p.10] The Big Seven hierarchy is a means by which to clarify boundaries, establish identity, and define order and deviance within the Usenet community.

The final community-wide strategy in Usenet is the network of sysops, or system operators. According to Hardy [1993]:

> "One essential element in the Usenet system is the system operator (sysop). In the traditional culture of the Net, the sysop has wide latitude to determine what activities she or he will permit on the machines over which she or he has authority. Since there is no central mechanism and no universally recognized standard for determining what is and is not a 'real' Usenet newsgroup, the system operator acts as a gatekeeper, determining which groups will be stored and forwarded from their host to other elements of the net to which they are connected. . . The sysop must weigh the desires of readers, the established customs and practices of the net and the possible consequences of conflicts due to controversial postings or allegations of censorship if a group is removed from the local feed. . . [pp. 3-4]

It is important to note, however, that while there is no 'official' consensus concerning what sorts of 'gatekeeping' measures are legitimate, there are a number of social controls which may be exerted over sysops, either by other sysops or by users of a particular sysop's system; the power structure is not a one-way mechanism. For example, in the news.admin.misc newsgroup, there is a periodic posting of 'sites honoring invalid newsgroups'. Sysops who find their sites listed in this posting are strongly encouraged through this sort of peer pressure to update their listings of newsgroups propagated in order to accommodate the wishes of the community.

The Site Administrator's Guide to Netiquette FAQ describes the basic rules which all sysops are expected to follow if they want to continue to be a part of the Usenet community. Because of the high degree of interconnectedness of newsgroups, sysops are expected to take their responsibilities seriously, particularly with regard to maintaining their newsgroup listings and propagating postings to the appropriate distributions.

It is the role of sysops to maintain the efficiency and legitimacy of the newsgroup hierarchy strategy. Sysops who do not perform this function adequately, who do not propagate the newsgroups that the majority of their readers want or who continue to propagate 'defunct' newsgroups, and who do not attempt to stop net-abuse, are labeled as renegades. "A renegade site frequently becomes the target of a boycott. This means that other sites will refuse to feed it articles, and will refuse to carry any of that site's posts. This situation is not too common. . ." [Usenet Site Administrator's Guide to Netiquette]

Within individual newsgroups, members use moderators, signature files and standardized message formats to establish order and maintain the norms and values of Usenet as a whole. The use of a moderator depends upon the group's specific needs. Among those that are likely to need moderators are: groups devoted to net news (e.g. news.lists, and news.announce.newusers); groups with a high volume of traffic; groups that have often been abused in the past; groups designed to serve also as direct feedback to an off-the-net group (e.g., the discussion in comp.std.mumps); and groups which are gatewayed into Usenet from an Internet mailing list. [List of Moderators for Usenet, 1995].

If one wants to post to a moderated newsgroup, one must mail the article to the appropriate 'submission' address; Usenet software does not allow direct posting to moderated groups, and it will not allow those unauthorized articles to be forwarded to other sites. More recent versions of news software automatically forward an attempted post to the moderator. Depending on the newsgroup, if the moderator decides that an article is inappropriate, he or she may return the article with a suggestion for other newsgroups in which to post it, and/or an explanation of why it is not appropriate for the moderated group.

Some Usenet researchers [MacKinnon 1995; and Kollock and Smith 1994] have suggested that this process of moderating a newsgroup is the equivalent of Hobbes' Leviathan, in which people give up part of their personal freedom to an authority in exchange for some measure of social order. While it appears to be true that, to a certain extent, individual newsgroup members rely on the judgment of the moderator to disseminate all relevant postings, this power structure is never one-way. Individuals are encouraged to contact moderators by email if they have questions or concerns relating to the moderator's judgments. If a satisfactory response is not forthcoming, the individual may resubmit the issue to a mailing list of moderators, who may then become involved in the situation.

Further, if members of a given newsgroup want to moderate the group, they must submit a Request For Discussion and Call For Votes through precisely the same channels as those used for newsgroup creation. In this way, all affected Usenetters have the opportunity to debate whether or not conditions in a particular group warrant such action. The decision to moderate a group is not taken lightly, nor is the potential for power within that position often abused. Moderators, like sysops, are accountable to the community as a whole because people have other channels of communication available to them (e.g. private email, unmoderated groups with similar topics, etc.). Moderators appear to pride themselves on being objective, on allowing all relevant views to be heard (even flame wars are often allowed), and on eliminating only those postings that are patently off-subject.

Regarding validation of personal identity, the signature file (or sig.) and the standardized message format are important strategies. Like the Undernet, Usenet allows for identification through both a nickname (or signature) and 'true-name' information. Nancy Baym's [1995] participant-observation of the rec.art.tv.soaps (r.a.t.s.) newsgroup reveals the obvious importance of names in establishing personal identity and a sense of group belonging in Usenet:

> "The obvious starting point in creating an identity is in the choice of a name. Anonymous CMC systems give people the chance to name themselves. Other systems attach real names, as do many Usenet newsreaders, but users may still sign their messages with names they have chosen for themselves. In r.a.t.s., people usually use their own names. Some take on nicknames, but most who use nicknames also promulgate their real names within the same message. . .participants on r.a.t.s. actively discourage anonymity. . . . Another means of creating identity, one related to naming, is the creation of a 'signature

file'. . . Signature files are attached automatically to the bottom of posts by the sender's newsreader. Signature files usually include a name and an email address. Other components often included are quotations, company disclaimers, and illustrations created by using punctuation marks and letters. . .Signature files, because they appear in the body of each post from a given sender, are one of the most immediate and visually forceful cues to identity." [pp. 154-156]

A few examples will serve to illustrate the ways in which signature files are used to establish a sense of identity for the self and for other newsgroup participants:

Example 1:

Damian Hammontree   damian@cthulu.med.jhu.edu
"A spokesman for the Lyon Group, producers of _Barney and Friends_, denied that Barney is an instrument of Satan." --the Advocate, spring 1994
Example 2:

Mark Isaak   "It is impossible for anyone to learn that
isaak@aurora.com   which he thinks he already knows." - Plutarch
Example 3:

Angie Tate
"There are two means of refuge from the miseries of life: music and cats."
       -- Albert Schweitzer
tateajt@gvltec.gvltec.edu
Site Coordinator, Greer Plaza Distance Learning Center
Greenville Technical College, Greenville, South Carolina

One can glean a fair amount of information about the individual who posted a given message, including true name, at least one email address, place of employment, and in some cases, philosophy of life. These are highly typical examples of signature files which, interestingly, provide the same basic types of information that Undernetters can obtain about one another, although through different means. In systems of communication with potential for 'anonymity', users seem to go out of their way to establish social order by being open about themselves; they readily offer personal information generally understood as relevant for the 'getting to know one another' process.

The second strategy that reveals information about individuals posting the messages is the standardized message format that all Usenet posts fit. Below is a series of three articles in the sequence in which they were posted to rec.pets.cats. This example displays nicely the issues relevant to both message format and group response to deviant behavior that violates group norms. Message header information reveals the subject and date of the post, who sent it, their organization, and newsgroups and reference headings. Consider the first post in the sequence:

Subject: Re: cat is still peeing!!!
Date: 21 Nov 1995 23:00:58 GMT
From: dacktyl@winternet.com (Farmer Fischer)
Organization: StarNet Communications, Inc
Newsgroups: alt.tasteless, rec.pets.cats, mn.general, alt.consumers.free-stuff
References: 1

Heather Thistle (hthistle@bbnplanet.com) wrote:
: Hi,

: I wrote a few weeks ago about my mom's 8 year old female persian. She
: was peeing all over the house. Mom brought her to the vets, and she
: checked out fine. Nor is there any behavioral habit suggestive of any
: moving of furniture etc. Does anyone have any advice? Should she bring
: Ashley to a different vet (this one charged a paw and a paw anyway!!!).
: Is there any way to stop her from peeing? And how can she get rid of the
: god awful smell that Ashley leaves behind?

: Any and all info will be very much appreciated!

An ice pick at the base of the skull jammed in real quick. No bleeding if you do it right. Then send the pussy to glory in a Gladbag[tm].

In this first post, the subject line contains a 're', meaning that this is not a new topic, but an ongoing 'thread' entitled 'cat is still peeing'. Information regarding the sender's email access and organization are disclosed, as is the fact that this response has been posted not only to rec.pets.cats, the group which started the thread, but also to alt.tasteless, mn.general, and alt.consumers.free-stuff. This is important to understand because those familiar with Usenet understand that any article posted to alt.tasteless is not likely to be one that is particularly amusing or helpful.[3]

Below, in the second post in the sequence, the 'from' and 'organization' lines are different, but still reveal the same types of information about the sender. The same newsgroups have been included in the cross-posting, which is another indication that this particular response will not be acceptable to the members of rec.pets.cats; as the last post will demonstrate, a member of the relevant newsgroup does not cross-post to the irrelevant groups.

Subject: Re: cat is still peeing!!!
Date: Wed, 22 Nov 1995 06:13:58 GMT
From: grante@reddwarf.rosemount.com (Grant Edwards)
Organization: Fisher-Rosemount, Rosemount Inc.
Newsgroups: alt.tasteless, rec.pets.cats, mn.general, alt.consumers.free-stuff
References: 1 , 2

Farmer Fischer (dacktyl@winternet.com) wrote:
: Heather Thistle (hthistle@bbnplanet.com) wrote:

: : I wrote a few weeks ago about my mom's 8 year old female persian. She
: : was peeing all over the house.

: : Any and all info will be very much appreciated!

: An ice pick at the base of the skull jammed in real quick. No bleeding if
: you do it right. Then send the pussy to glory in a Gladbag[tm].

Bravo!
A brilliant play on words, Steve.
Pissing... pithing... get it folks?!
Pithing pissing pussy prevents pungent puddles; pictures posted post-haste.
--
Grant Edwards | Microsoft isn't the | Yow! Do you have exactly
Rosemount Inc. | answer. Microsoft | what I want in a plaid

| is the question, and | poindexter bar bat??
grante@rosemount.com | the answer is no. |

Finally, in the last post, an individual who is a regular on rec.pets.cats responds in an attempt to reestablish group order. She attempts to define Farmer Fischer as deviant and to account for both his actions and the fact that his access provider allows such behavior. She does not re-post his material, however, because to do so would perpetuate the problem. If group members are unaware of this discussion, they may refer to any or all of the three references listed in the header information.

Subject: Re: cat is still peeing!!!
Date: 22 Nov 1995 10:48:34 GMT
From: capella@winternet.com (Camille Klein)
Organization: StarNet Communications, Inc
Newsgroups: rec.pets.cats
References: 1 , 2 , 3

sigh

Please ignore Farmer Fischer. He's just some git from Minneapolis who
has nothing better to do with his time than troll for flames by posting
the most gross and rude crap he can think of to whatever newsgroup.
Complaining to his ISP will do no good, as Winternet does believe in free
speech and will not censor any users.

Please learn to use a killfile, and put the Farmer in it. Either that,
or skip over his postings altogether if you do not agree with them. Thanks.

--Camille Klein, a Winternet user and no fan of Farmer Fischer.

These Usenet conventions--FAQs, newsgroup creation strategies, the Big Seven hierarchy, sysops, moderators, signature files and standardized message formats--are all instrumental to the ordered functioning of the Usenet community. They provide for the individual user a sense that there is a group; a group with established norms, values and belief systems, through which he or she may derive a sense of identity and belonging.

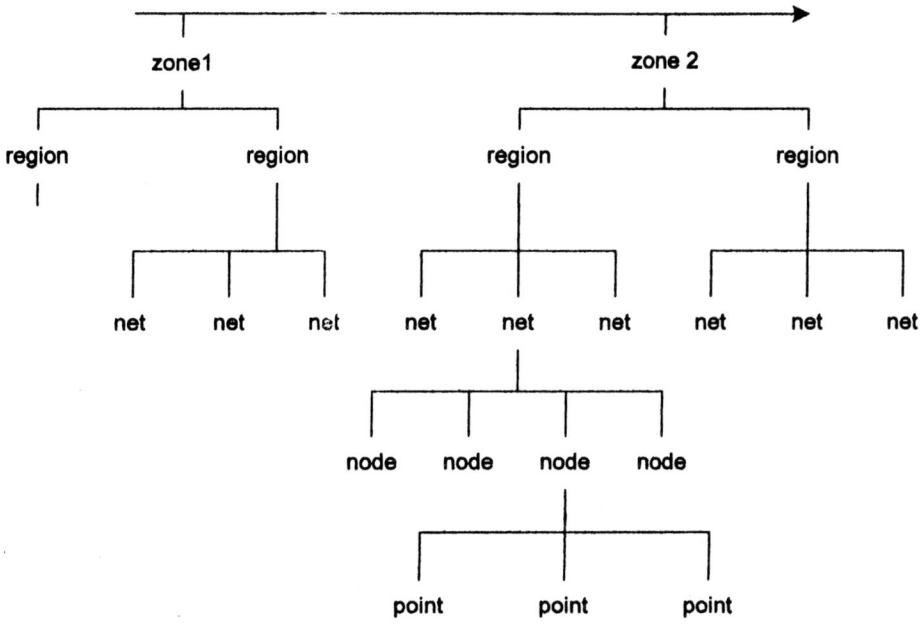

Figure 2. Fidonet Connectivity

THE FIDONET COMMUNITY

Interestingly, Fidonet bears resemblance to both the Undernet and Usenet in terms of its technical and social structures and its atmosphere. Like these communities, Fidonet has both community-wide and within-conference social institutions that afford it a sense of unity and order. The principal community-wide strategies are the BBS node system, *Policy4*, and FidoNews, the official newsletter of Fidonet. Like the Undernet's community-wide institutions, these strategies are vital to the continued technical stability of Fidonet and even have a broad influence over the community's social order, yet they have relatively little influence over the day-to-day activities of the BBSs and the conference areas they transmit.

The estimated 30,000 BBSs in the Fidonet system are linked together in what is called an hierarchical node system, as the figure above illustrates. This design, which was formalized with the passage of *Policy4*, was originally developed by Tom Jennings and other Fidonet leaders as a strategy for simultaneously decentralizing the power structure of the community and establishing order in terms of which BBSs are legitimate community members. BBS organizers who do not go through the formal procedure of getting an official node number are not listed in the official NodeList. It is this NodeList that defines the boundaries of the Fidonet community.

This new node system strategy was generally accepted by the community in 1985 in the form of Policy1. Policy1 has since evolved into *Policy4*, which formalized not only the BBS node system and NodeList procedures, but a variety of other technical and social policies. According to Bush [1993]:

"In 1985, the first Fidonet policy document was published. It concerned itself almost entirely with technical procedural issues. It required a capability to send and receive email, defined the 'national mail hour' as mandatory, delineated roles of the local network hubs and nodelist coordinators, and stated simple restrictions on routing of traffic through unsuspecting nodes. In addition, it stated two social rules, a proscription against use of the network for illegal purposes (e.g. pirated software) and a statement of Fidonet's basic social guideline, 'do not be excessively annoying and do not become excessively annoyed.' ...

The first written policy was published and adopted by informal consent. Subsequently, three revisions of Fidonet policy have been written and made operational by various, but less democratic, procedures. The current document, *Policy4*, was written by the regional nodelist coordinators, and has a large amount of social and political content enshrining a hierarchy of coordinators: An International Coordinator (IC), a Zone Coordinator (ZC) in each continent, Regional Coordinators (RCs) in subdivisions of the continents, usually countries, and a Network Coordinator (NC) for each local network. As it was written by the self-anointed RCs, ZCs and the IC are elected by the RCs and NCs are appointed by the RCs. Although the document has caused considerable acrimony and is large and complex, it contains many useful operational guidelines, so is generally observed." [pp. 6-7]

Besides the organizational hierarchy, *Policy4* also 'formalized' what had already been generally accepted by community members as a whole as the 'national mail hour'. Fidonet calls were originally set up to dial automatically during the middle of the night, when long distance phone rates were at their lowest. Over time, as the number of nodes in the community grew, it became increasingly critical to time these calls successfully. Thus, the general Fidonet community policy was established which requires the sysops (who run the BBSs) to set up those BBSs to receive calls at the appropriate time during the night: "Zone Mail Hour is the heart of Fidonet, as this is when network mail is passed between systems. Any system which wishes to be a part of Fidonet must be able to receive mail during this time using the protocol defined in the current Fidonet Technical Standards Committee publication. . ."[*Policy4*, 7]

The importance of the national mail hour for sustaining the sense of community for the 30,000 BBSs involved has become even more evident over time. Even though advances in Fidonet technology now make it possible to intermix mail transfer with BBS access, community members have elected to continue the national mail hour as a time when they all 'come together', in a sense, and accomplish a community-wide goal; they continue the ritual, even though it is technically possible to do away with it.

Sysops in the Fidonet system serve the same functions as sysops in the Usenet system, but often times are not as highly regarded. Fidonet was born of a desire on the part of a few individuals to 'hack'; to develop CMC as a hobby. Over time, Fidonet culture emerged as a myriad of individuals, varying widely in interests, temperaments and abilities, set up their own BBSs and became part of the NodeList. So, though the culture and sense of camaraderie among Fidonet sysops exists because of this history, there also exists among some sysops (and many regular users) a sense that a fair number of Fidonet sysops are "twits" or "clueless".

Fidonet sysops are those people who choose to set up a BBS on their own computer. They may design their BBS in any way they please, meaning they may set up local conferences, choose which Echo conferences to receive, choose whether or not to receive games, and even whether to bear the expense of a high speed phone line that will allow callers to chat in real time (in a way similar to IRC channels). Further, the sysops decide what the log-on registration requirements for their BBSs will be, whether or not to carry 'adult' sections in their BBSs, and whether or not people who use their BBSs will be allowed to use nicks or true names only. In general, then, sysops have wide latitude in designing and maintaining their Fidonet BBSs. However, there are certain responsibilities that sysops must live up to regarding familiarity with policy, encryption, review and alteration of netmail, and the behavior of those users logging into the Fidonet system via their BBSs.

Because BBSs come and go, the Fidonet community depends upon a fairly steady influx of new BBSs (and new sysops) to sustain itself. These sysops come from all walks of life, and thus contribute to the diversity and vitality of the Fidonet culture. Being a Fidonet sysop affords one access to a variety of other community members with whom one can discuss and debate almost any topic; it allows one to gain a sense of membership. In return, sysops are expected to follow a few basic rules that allow for the continued orderly flow of interaction which is the essence of Fidonet.

Ironically, while the BBS node system, the national mail hour and *Policy4* are in one sense the 'glue' that holds the Fidonet community together, in another sense these strategies are relatively unimportant in the day-to-day workings of community life; on one level, they are essential, on another, barely noticeable. Where the Undernet has its Boards of Coordinators and User Committee to maintain the technical infrastructure and oversee political issues, Fidonet has the hierarchical Coordinator's system and *Policy4*. Yet, just as anyone can join the Undernet and create his or her own channel, anyone can join Fidonet by simply taking the time to set up a computer as a node and calling for an official node number. In contrast to the Usenet newsgroup creation process, neither the Undernet channel creation nor Fidonet node creation involve any sort of community-wide decision process.

Additionally, the NodeList containing each BBS's phone number is distributed worldwide once a week, and is easily accessible for any sysop who chooses to download it. This means that any node anywhere in the world may talk to another node without being required to jump through any technical or political hurdles in order to do so. This is why, even though *Policy4* is hated by many Fidonetters, they allow it to go relatively unchallenged--it is easy enough to 'get around' the 'official' structure.

Finally, just as the purpose of Undernet interaction is to 'hang out', Fidonet interaction most of the time has a fairly 'laid back' feel to it. Though one enters specific conference areas to discuss specific topics, newbie questions and a fair amount of "off topic" discussion are tolerated (depending upon the conference). Fidonetters seem perpetually amazed that the system even works; they count it as a pleasurable way to spend time. Thus, 'serious' rules and policies and whatnot, though generally adhered to, are also generally considered irrelevant.

Before moving to a discussion of within-conference institutions and groups, one final community-wide institution is worth mentioning--FidoNews. FidoNews is the official newsletter of the Fidonet community, and is a key means by which Fidonetters, especially the sysops, stay in contact with the rest of the community. Each edition contains an editorial, one or more feature articles, and discussions of computer programs, problems, services, bugs and announcements. Each edition is also marked by the official community emblem, the Fido dog with a disk in his mouth. Sysops are responsible for downloading FidoNews from a node to which it is regularly distributed, and for making it available to individual users, if they so choose.

Echo conference areas are Fidonet's equivalent to Usenet's newsgroup interactions and the Undernet's channel interactions; conferences are the places in which an indefinite number of people may 'meet' and discuss the topic at hand. And like channel and newsgroup members, members of Echo conference areas have designed their own particular problem-solving strategies to meet their particular needs.

Fidonet Echo conferences are an odd combination of the goals of Usenet newsgroups and the atmosphere of Undernet channels. Each conference is named, providing a frame through which participants may define interactions; conference names serve to define the topical versus off-topic posts, appropriate and inappropriate tones, acceptable comments and flames. However, many conference areas lack the feel of 'seriousness' found in many Usenet newsgroups. While one is expected to stay basically on-topic, the purpose of 'hanging out' is also accepted. Additionally, conference areas have in common with Undernet channels a feeling that it is acceptable to ask newbie-type questions. For these reasons, the Fidonet community has not developed the community-wide FAQ as a socialization strategy. One is expected to learn the ropes through interaction in particular conference areas, just as one does in particular Undernet channels.

However, members of individual conferences have developed documents known as conference Rules, and as one becomes familiar with Fidonet, one learns that most, if not all, conferences have Rules. Because of this, it is commonly expected that newbies will take the time to read the relevant conference Rules before actively participating in any discussions. Failure to do so often results in a reprimand. These conference Rules serve the same purpose as Usenet newsgroup-specific FAQs; they follow the general community guidelines, but are tailored to the specific conference atmosphere.

These conference atmospheres range somewhere between the purely 'chat'-oriented atmosphere on one end and the purely topic-oriented atmosphere on the other.

Below are some sample conference Rules, which generally revolve around a few common issues considered essential to social order--advertising, the use of nicks, appropriate tone of posts, treatment of newbies, and so forth. The examples begin with one that falls close to the purely topic-oriented end of the continuum and conclude with one (from the Mindless Chatter and Drivel Conference) that falls toward the purely chat-oriented end of the continuum:

Example #1:

Area: FIDO - Psychology
Msg#: 3
From: Bob Johnstone
To: All
Subj: RULES

[PUBLIC PSYCHOLOGY]==========================[94.10.10]

General information for EVERYONE:

Echomail, is a group discussion of topics. Assume it is an open topic just AS IF ADDRESSED TO ALL: and not a message only for an individual. Everyone is welcome to participate in any topic, at any time, without asking "permission." Personal or private messages belong in NETMAIL.

But all messages must be about Psychological Support. What you do, or ways you are able to cope and function. Ideas that others can use to help themselves, without any criticism and only supportive messages or requests for help are allowed.

As the Rules apply to ALL, these rules are considered a -warning- to all and adults don't need private or personal warnings to abide by rules.

OFF TOPIC MESSAGES increase phone bills for -hundreds- of SysOps, please be considerate of those paying long distance rates, stay on topic.

LEGALITIES: Sciences of the Mind - Self Development Center owns the Compilation copyright for this "Electronic Newsletter," about Psychology.

Information Published in this Newsletter - "PUBLIC PSYCHOLOGY," just as articles in other publications remain the property of the author. Every message must be considered copyrighted, and used only with the specific permission of the author.

The Mental Processes called "Conflict Resolution" and "RELAX TO THE MAX" are copyrighted. They are also a Trade Mark of Sciences of the Mind.

The Moderator of the Conference since 1978 has been Bob Johnstone. For information NETMAIL 1:10/25, or the BBS 714-525-1706 & 1708.

This Conference is a service to those who need emotional or mental and Psychological help for life's experiences. You can improve the quality of life or reduce and even eliminate stress or the effect of memories. Criticism or attacks upon individuals are not allowed.

HOW do you deal with life effectively? Please give as many details as possible or the sources, like books and therapy which helped you.

HANDLES: Are encouraged, if allowed where you access the conference. But, please use one, that sounds or looks like a real name. As most individuals prefer to talk to a "REAL PERSON" and weird Handles, do not seem to encourage communication. But, be aware there are Native Americans and others who use traditional names and respect that right.

Advertising is allowed, if appropriate or consistent with the topic. ONLY advertising which will benefit users psychologically is allowed, and it requires PRIOR permission from the Moderator if you have doubts. For a Review of advertising, send NETMAIL to Bob Johnstone, 1:10/25

For a Review of tapes, scripts or books please send a copy TO: Dr Robert Johnstone, P.O. Box 5122, Anaheim, Ca 92804. Please, Include a script tapes so I don't have to transcribe them.

FREE HELP with problems that cause stress,

Bob Johnstone, 1:10/25,
Moderator STRESS_MGMT
Moderator PUBLIC_PSHCH
BBS: 714-525-1706 & 1708
VOICE ONLY: 714-525-1036

Example #2:

Area: FIDO - US and World News
Msg#: 36
From: Joe DeLassus
To: All
Subj: ANEWS rules & guidelines

The ANEWS echo is a forum for news articles and stories. It is meant to present news stories and points of view that are not covered by the corporate-controlled, mainstream mass media. This echo is not meant to be a particularly well-rounded or so-called objective news echo.

It is advised that readers compare ANEWS stories to stories found in the mass media to detect various news angles, slants, and omissions (from both/all sources of news). An impossible task, of course, but one should vary their reading diet.

If you would like to routinely enter news articles or stories into ANEWS, please get permission to post the articles from the source/publisher that you receive the news articles from, and then please send a message to the moderator of ANEWS informing him of what sort of news you plan to post.

If you would like to intermittently post news items or comments about the news to ANEWS feel free to do so, but please give proper credit if you are quoting or referencing a source and indicate what parts you are quoting versus your own analysis/commentary. And, please try to keep individual posts down to 8K (approx. 115 lines) since some systems do not process messages larger than this.

Understanding Community Structure: Social Institutions and Groups

The individual sysop can, at his or her own discretion, make the ANEWS echo read-only or not depending on the Sysop's preference. However, messages marked 'private' should not be posted in this echo. Replies to specific news items appearing in ANEWS or general discussion is not recommended or encouraged.

The ANEWS echo can be freely passed to other nodes as long as the echo is passed in full, with no messages having been deleted and/or altered (except alterations done by normal echomail processing utilities).

This echo is available on the Fidonet Zone 1 backbone, but please spread the echo around to whatever node desires it to make it as widespread as possible.

As far as specifics about the rules, the moderator has the final say in all matters and policy concerning this echo.

Generally annoying behavior and personal attacks on other participants, readers and/or posted news items in ANEWS are grounds for expulsion from the echo.

If you have any questions, ask here in ANEWS or via netmail to me, the moderator, Joe DeLassus at Fidonet 1:100/355 or assistant moderator, Scott Parks at 1:343/70

Example #3:

Area: Chatter
Msg#: 21614
From: Timothy H. Willis
To: All
Subj: Da Rulz<tm> one last time
-!!- insert Twilight Zone theme music here -!!-

>Mindless Chatter<tm>: The rubber room of Fidonet<tm>...

 To: All Newbies: Pay close attention, now... a quiz follows...
 >Mindless Chatter & Drivel: RULZ<tm>, warning(s), & game pieces...<

***** Dis is a rendition of DA RULZ<tm>, brought to all youse witless Newbies today, 27 December 219 AL, courtesy of Timothy H. Willis... sponsored by:
Skidmark Enterprises<r>. READ!!! LEARN!!! WET YOURSELF!!! *****

>Mindless Chatter & Drivel: Arguably, the WORST echo in Fidonet<tm>...

(Haven't seen 'em in a while, so I guess it's MY turn to post 'em. <sigh>)

Area: CHATTER

In this conference, EVERYONE moderates. It's in the rules you so obviously did not read. Here's your own copy:

Moderators: Martin Goldberg, 1:124/9005.221@fidonet.org (and anybody else that has proven they know how to post "on-topic" drivel) !!-
THE RULES FOR MINDLESS CHATTER.. Abridged Version 9.2.1, for 1995!

... Mindless Chatter-!-This ain't no "CHAT with your friends" echo.

AREA CHATTER
TITLE Mindless Chatter Conference
DESC A conference devoted to babbling on and on about utter nonsense. For those in the United States, Canada and parts unknown. People who post serious topics, personal messages, BBS ads, or advertising of ANY KIND will be severely chastised by everybody in the whole blamed echo. Repeat offenders may be added to our collective TWIT list, and/or suffer invasion by armies of enraged dwarfs armed with .50 cal Aardvarks. You HAVE been warned!

Messages are light-hearted, with no chance of accidental seriousness.

A perfect place for stressed-out persons to find fast relief. Hershey Bars allowed. Aardvarks allowed. Cat jokes allowed. Suspend your reality at the

Understanding Community Structure: Social Institutions and Groups

door and be ready to make fun of yourself. No gross profanity allowed, unless you back it up with math proofs and brute force.

... If you want to <gasp> "CHAT", try Interuser, or Coffee-Klatch...

An Explanation Of Implied Rules So That Even The Most Ignorant Person Has An Opportunity To Learn Them Painlessly Before We Kick Your Butt!

1) NO BBS ADVERTISEMENTS! THIS MEANS YOU. I DON'T CARE HOW PROUD OF YOUR STUPID BBS YOU ARE! (After all, if it's carrying MINDLESS CHATTER, how good can it BE?) Posting a BBS ad is grounds for getting your feed cut. If I'm in a really VILE mood, I won't cut your feed, but I *WILL* put your name on a mailing list for pictures of Rush Limbaugh wearing skimpy women's underwear. Enough said, yes?

2) I've just been informed by The Powers That Be, who not only live in my apartment, but also read over my shoulder, that Rule #1 was discriminatory. I apologize and offer this rule to rectify that situation. Therefore, NO ADVERTISEMENTS WHATSOEVER ARE ALLOWED! The same penalties as above apply.

3) This is a nationally distributed, public echo. It reaches thousands of readers at hundreds of BBS's, and each and every one of them has the right and potential ability to read any bloody message they want to in this echo. So, DON'T BRING YOUR PRIVATE CONVERSATIONS HERE AND EXPECT

THEM TO STAY THAT WAY! Private conversations belong in NetMail. Talk to your Sysop for more information.

4) Last time I looked, there were about five hundred other Echos available to the backbone, with set topics ranging from Abbots to Zymurgy. If you want serious discussion, find another echo. Chances are that there are SEVERAL that will serve your needs.

5) There are two topics of discussion that, while arguably drivel, will NOT be tolerated in MINDLESS CHATTER: religion and politics. Proselytizing and campaigning will not only be met with verbal firepower, but also with parrot droppings, turnip greens, and Martin's mother-in-law's meatloaf. You'll be begging us to cut your feed by the time she gets to the grape jelly potato latkes.

6) This is NOT the FLAME echo. Mild, gentle flames are acceptable when a newbie who didn't bother to read these rules first steps in and does something wrong. However, protracted abuse is not acceptable under any circumstance. If you MUST blow off steam, please do so in NetMail. (Besides, if you're going to act like an infantile, screaming jerk, do you REALLY want to do it in front of EVERYONE?) Well, just ignore that, we know it's fun to do it in front of everyone...

7) Don't even THINK about applying for Membership unless you are willing to send $127.50 for Drivelling Papers, Phunny Hats, Decoder Rings and any other souvenirs we might dream up. And don't be surprised if you don't RECEIVE any of the aforementioned items. We make absolutely NO CLAIMS about our honesty here.. only that you'll regret your visit.

8) Be it hereby known that we reserve the right to poke fun at ourselves, Catholics, Protestants, JWs, Mormons, Jews, Muslims, homosexuals, heterosexuals, androgenous people, asexuals, celibates, televangelists, witches, cats, dogs, squirrels, fish, waterfowl, boars, ethnic people, Gregorian Chanters, stone masons, gondaloons, bedoobies, passflooses, basselopes, Michael Ching-a-ling, The Winslow, Anita Chircop, any and ALL Newbies 'n Clueless Syssiops, Scott Heald (Clueless of the Month/May), Josh (huh-huh-huh) Hoffman, Nicole (da blonde) Sherman, Ben-Wa gerbils, Canadian Moose-lovers, and anything else our twisted minds can devise.

9) All things being equal, The Moderator shall reserve the power to slap you around suddenly and for no apparent reason.. just for the ugly thrill of it. When you get to be Moderator, YOU make the rules.

10) Sigh. Okay, Phil, this one's for you. NO PUFTAS! (Don't ask. *I* don't know, either.)

>Mindless Chatter: Built upon the crushed corpses of witless newbies...

If you decide to stay, be sure to take your brain out of gear first, and check your reality at the door... Most newcomers get chastised the first time they post an "off-topic" message, and it gets progressively more abusive until they either leave, suffer a feed cut, or catch on and start posting drivel...

Besides a general understanding of the rules of various conferences, one can glean two other important pieces of information from the documents above. First, one can get a general sense of the atmosphere of a conference before one begins participating actively in the conversations. This allows the newbie to avoid making many mistakes that he or she might otherwise fall victim to. The first message a newbie should look for in his or her first group of downloaded messages is the one titled "The Rules".

The second important piece of information picked up through reading conference Rules is that every conference has at least one moderator (unlike Usenet, in which only a portion of the newsgroups are moderated), and that these moderators are an important strategy for preserving the order and identity of the conference. While Usenet newsgroups have to go through a formal voting process before they can be moderated, this is not the case with Fidonet conference moderators.

Echo conference moderators are elected by those who participate in the conference, but fill a role that is more similar to the Undernet channel op. They do not screen messages in advance. Rather, they actively monitor the messages flowing through the conference and warn those individuals who are getting 'out of line'. If the deviant behavior continues, the moderator has the authority to eliminate that individual's ability to participate in the conference; the moderator may 'cut his or her feed' (just as the channel op in Undernet can kick/ban). Additionally, conference moderators often take on the role of welcoming the newbies to the conference and answering any questions they may have.

Besides conference Rules and moderators, another key strategy for maintaining community order and identity is the tagline. Taglines are the equivalent of Usenet signature files in that they are created to make statements about one's self. Because the message headers and closers in Echo conferences contain 'essential' identity-documenting information for the other (such as true name, the name of the BBS on which the message originated, netmail address, and so forth) the sole purpose of the tagline is to allow a bit of one's 'personality' to be displayed to the other. Through the use of taglines, members attempt to make statements about and make assessments of their attitudes, intelligence and wit. In general, compared to Usenet signature files, taglines tend to be more colorful, more clever, and a bit less serious or philosophical (thus reflecting the overall difference in atmospheres between Fidonet and Usenet). In fact, taglines may even become the subject of an ongoing thread within conferences, and are frequently "swiped" and collected.[4]

Below are some sample taglines from a variety of conferences, through which one can get a general understanding of their role in establishing the identity of both the owner and the group, or audience, toward which they are directed:

Example #1 (from the Consumer Report conference, subject: buying a computer):

...Fer sell cheep: IBM Spel chekker. Wurks grate.

Example #2 (from the Mindless Chatter and Drivel conference, subject: newbie flame):

...Even my mouse has big balls.

Example #3 (same conference and subject as example #2):

...Mindless Chatter: Forget all questions before entering, inquire within.

Example #4 (from the UFO Discussion and Science Thought conference, subject: evil aliens):

...Are you coming quietly, or do I need earplugs?

Example #5 (from the Mindless Chatter and Drivel conference, subject: newbie):

...COMPUTER.COM installed. SEXLIFE.EXE removed from memory.

Example #6 (from the X Files conference, subject: episode):

...You never draw my bath. (Mulder to Scully).

Such are the key institutions and groups in Fidonet life. The BBS node system, *Policy4*, FidoNews, Echo conferences (and their accompanying Rules), moderators and taglines are all essential to the orderly functioning of the Fidonet community. It is through these accepted strategies that community members establish their personal identities, values and systems of belief. Regardless of whether one can define such strategies as 'formal' in the traditional sense, community members behave 'as if' these rules and regulations constrain their behavior. As in daily life in face-to-face communities, what holds IRC, Usenet and Fidonet groups together is that they agree to conform.

ENDNOTES

[1] The source for this information is an expert on IRC who started and maintains his or her own channel in the Undernet based upon experience in EFNet. This information was obtained in private email correspondence. Additionally, the individual's first language is not English, and the text has been left in its original grammatical format to better convey the personality of the individual.

[2] 'Spam' is the common term for excessive cross posting, in which a user posts a message to too many groups and to groups in which the message is irrelevant to the topic at hand. There are specific rules within Usenet for defining potential spam as 'actual' spam, in which case the use of a cancel message is warranted.

[3] In addition to marking previously posted materials with an ":", such materials may also be marked by an > at the beginning of each line of text. For an article embedded within that second layer, each line of text would be marked with a double >>, and so on. This system of text marking is extremely important for following the thread of conversation, and poorly edited quotes can lead to a great deal of confusion over who said what.

[4] "Swiping" a tagline consists of copying someone else's tagline when one really likes it, but freely admitting that one did not come up with it on one's own. It is not literally removed or taken from the other person's signature. Regarding tagline collections, conference members frequently swipe taglines they see elsewhere, but which are related to the conference topic.

CHAPTER 5

SOCIALIZATION AND PERSONAL IDENTITY: THE CREATION OF AN ON-LINE SELF

How do on-line community members encourage newcomers to abide by established group norms? How do they use the institutional strategies they have devised on a day-to-day basis to sustain social order? In a nutshell, they accomplish this in the same manner as do people in face-to-face communities; through language, through talk.

As discussed in Chapter One, research for this book was guided by the conceptual framework of symbolic interaction; the study of the making of meaningful behavior. In general, symbolic interaction theory states that human beings have the capacity for thought; this capacity is shaped by social interaction, in which people learn meanings and symbols that allow them to develop that capacity. Further, these meanings and symbols allow people to carry on action and interaction, and people can modify or alter meanings and symbols on the basis of their interpretation, or definition, of the situation. People can make these modifications because they have selves. That is, they can interact with themselves, examine courses of action and the advantages and disadvantages of each, and then choose one course of action over another. Finally, these intertwined patterns of action and interaction are what constitute groups and societies.

Two implications of symbolic interactionist theory are of critical importance; the self as process (rather than entity), and the role of language (symbol systems) in the simultaneous creation of both the social self (socialization) and the social order. Through interaction, one develops a self; the capacity to observe, respond to, and direct one's own behavior (reflexivity). That the self continuously arises out of interaction implies both interaction among individuals and a prior existence of the group(s) within which the self emerges. This process of self development, often called socialization, is a life-long procedure; the self is a process, not an entity, and is neither the product of impinging stimuli, nor a reflection of an overarching and overwhelming cultural system, nor an organism driven by and essentially determined by internal mechanisms. [Hewitt 1984] As Goffman [1959] aptly summarizes, the self is not an organic thing that has a specific location. "In analyzing the self then we are drawn from its possessor. . .for he and his body merely provide the peg on which something of collaborative manufacture will be hung for a time. . ." [p. 253] The self is not the possession of the actor, but the product of interaction within a scene or frame.

And what specifically constitutes this scene or frame? According to Blumer [1969], "Symbolic interaction involves interpretation, or ascertaining the meaning of the actions or remarks of the other person, and definition, or conveying indications to another person as to how he is to act." [p. 66] In other words, the self and social order are simultaneously framed through the symbol system of language. Language serves as the basis for everyday social life; it is the cohesive factor for human groups, because groups are symbolic, rather than physical, phenomena. Through the use of language, individuals create their selves and their reality; they indicate to themselves and each other what their responses to objects will be, and thus what the meanings of those objects are. [Lauer and Handel 1983]

Implicit in the process of reality-construction, language serves as a means of socialization for new selves. Language allows for the transition from biological entity to human self through role-taking, and this role-taking process is the basis for society. Cooperative processes are necessary for the maintenance of social order, and cooperative processes can only occur to the extent that individual members are able to apprehend the general attitude and, therefore, predict the behavior of other members of the society:

"From the standpoint of its individual members, a society is a thing with an existence independent of themselves, even though its continued being depends very much on them and their behavior. A society, in short, is an object toward which its members act, and to a great extent the fact of social order is simply the fact that people act toward, and so constitute, this object in a stable, orderly fashion. What is true of society as a whole also is true of the smaller groups, organizations, communities, institutions and other units that make it up." [Hewitt 1984, 188]

The primary means by which members 'act toward' and 'constitute' community, social order and the self is through talk. As Hewitt points out:

> "This points to the crucial importance of that form of conduct we call 'talking'. For on whatever occasions it takes place, talk is a primary means by which people sustain the world of objects in which they live, particularly abstract objects such as goals, rules, ways of life, institutions, groups, communities and society itself. Talk thrives best when conditions are in some way problematic. People account for their conduct when others define it as problematic or when they think they are likely to do so. Such forms of talk as accounts and disclaimers themselves make an important contribution to social order. By calling attention to rules, norms, and expectations, these forms of talk allow homage to be paid to the usual, regular, and typical ways people are expected to behave. They call attention to social order, even though the circumstances in which they arise involve apparent failures of that order."[p. 189-190]

As one can see from the descriptions that follow, this basic process of socialization occurs in on-line communities in much the same fashion as it does in more traditional communities. It occurs through talk.

THE INTERNET RELAY CHAT COMMUNITY

The key institutions and groups within IRC, as discussed in Chapter Four, are bots, public and private channels, nicks, ops (and newbies) and the various "/commands". This chapter, however, focuses on two specific points with regard to the process of socialization and the presentation of self within IRC; the uses of public and private channels and the "/commands", each of which is critical for socialization and definition of the situation, for presentation of self and social identification, for the formation of significant others, and for resolving ambiguity and achieving reciprocation regarding communication and the self.

Chapter Three described the general culture of the Undernet as the Third Place. For the most part, discussion is relatively light-hearted and friendly. All channel members are encouraged to join in the conversation, provided that they follow the basic rules of conduct; the basic norm is that one should be 'friendly'. However, the medium itself and the culture of IRC demand that one be speedy and witty in ones responses to others, because interaction in IRC is, for the most part, a stream-of-consciousness dialogue in which conversations move very quickly from one topic to another. Additionally, in any channel with more than four or five people, there are likely to be several conversations in progress at any given time. People may engage in multiple conversations or just one, and on occasion these conversations may merge.

Within this general cultural framework, community members, especially the regulars, use public channels to socialize newbies to the rules. In most cases, it is fairly obvious who the newbies in a channel are, either because their nick is not recognized, or they commit an error in their presentation of self that indicates to the regulars a newbie status. Because of the easy-going nature of IRC interaction, newbie-ness is not considered a problem (deviant) unless it is defined by the regulars as being intentional.[1] For example, in the dialogue below, Dorothy is a newbie to channel #25plus, and begins to learn the ropes:

[DISP-#25plus][SERVER]   dorothy!~dorothy@204.50.107.20 has joined this channel
[DISP-#25plus][kippers]   hi dorothy
[DISP-#25plus][kippers]   kippers welcomes dorothy to the friendly channel ... :)
[DISP-#25plus][dorothy]   thanks for the welcome
[DISP-#25plus][kippers]   you're welcome
[DISP-#25plus][ACTION]   salt waves at dorothy ~
[DISP-#25plus][kippers]   where are you dorothy?
[DISP-#25plus][ACTION]   OrbWeaver says hello also to dorothy
[DISP-#25plus][dorothy]   I'm from Canada
[DISP-#25plus][salt]   in BC maybe
[DISP-#25plus][dorothy]   i'm in Ontario
[DISP-#25plus][salt]   dorothy, are you a little new on this channel?
[DISP-#25plus][dorothy]   how did you guess? Did my newbieness give me away?
[DISP-#25plus][ACTION]   OrbWeaver did not detect overt newbieness.
[DISP-#25plus][dorothy]   what's the channel about?
[DISP-#25plus][salt]   dorothy...you use capital letters....a newbie-sign :)

[DISP-#25plus][dorothy]   thanks, i'll stop doing that
[DISP-#25plus][ACTION]   salt smiles
[DISP-#25plus][ACTION]   Marcia_Br giggles
[DISP-#25plus][ACTION]   OrbWeaver uses capital letters sometimes .. afraid his 6th grade teacher will return from the dead to correct him.
[DISP-#25plus][ACTION]   dorothy laughs
[DISP-#25plus][salt]   dorothy, if you are curious about the channel and the people...look at our home page
[DISP-#25plus][ACTION]   OrbWeaver tries to use lower case .. a unix thing. oops, Unix.
[DISP-#25plus][salt]   hehe orb
[DISP-#25plus][salt]   it's a little guide into the wonderful world of channel 25plus :-D

In this example, Dorothy is greeted warmly, but is not recognized by the regulars and is thus questioned about where she lives and whether she is new to the channel. Such questions are commonly put to newbies; they are asked where they live, what they do for a living and so forth. While such questions are used to 'get a feel' for what kind of person the newbie is, they are rarely asked in a 'pushy' manner, and the newbies rarely take offense to them. Dorothy admits to her newbie-ness and responds well to the suggestion that she alter her interaction style, or presentation of self, in order to give the appearance of a regular. In terms of defining the situation, it was important that Dorothy acknowledge newbie-ness and accept help if she intended to avoid being labeled a deviant.

Contrast this example with the one below, in which 'junaid' violates the all-caps norm but refuses to pay attention to the suggestion that this makes him 'weird'. Initially, the channel members continue to ignore junaid, in the hopes that he will correct his behavior. Eventually, both kokopelli and n22 attempt to inform junaid that he should not be using all-caps to speak because it is the equivalent of shouting in the channel. When these polite suggestions do not work, Romario suggests that junaid change his behavior in a bit more forceful manner. This is typical of how people new to a channel are socialized to the rules. When the rule infraction appears to be unintentional on the part of the newbie, the dialogue by the regulars begin mildly, with suggestions that the newbie may want to change his or her behavior. If this sort of negotiation does not work, the regulars usually attempt increasingly harsh means of negotiation. It is often the case that, if the newbie persists in refusing to negotiate the situation in a manner that the regulars feel is acceptable, the newbie will be kicked from the channel and possibly banned. Eventually, as people in the channel continued to ignore junaid, he took Romario's suggestion and began typing in small letters instead. Shortly thereafter, several members of the channel remarked that he had 'changed' and began an open and friendly dialogue with him:

[DISP-#chatzone][junaid]   HI EVERYONE ANY BODY WANTS TO TALK PLEASE URGENT
[DISP-#chatzone][SideSwipe] what junaid
[DISP-#chatzone][kokopelli] hi all
[DISP-#chatzone][Butch]   hi, i've just joined
[DISP-#chatzone][Nessie]   hi kokopelli..
[DISP-#chatzone][junaid]   I AM GREAT
[DISP-#chatzone][kokopelli] why junaid?
[DISP-#chatzone][SERVER]   n22!r00t@etc.tcl.tec.sc.us has joined this channel
[DISP-#chatzone][Poison] hi kokopelli!:)
[DISP-#chatzone][junaid]   KOKOPELLI WHO ARE YOU, MY FRIEND
[DISP-#chatzone][Brat]   Does anyone want to talk?
[DISP-#chatzone][Chatbot-MODE] Has changed n22's mode to +o
[DISP-#chatzone][leXie]   i always want to talk, brat.
[DISP-#chatzone][Poison]   hi Brat!
[DISP-#chatzone][Marie]   hi Brat
[DISP-#chatzone][kokopelli] junaid why are you yelling?
[DISP-#chatzone][n22]   heya brat. how ya doin? :-)
[DISP-#chatzone][Nessie]   yeah Brat, tell me ya problem..
[DISP-#chatzone][Emu]   ill talk brat
[DISP-#chatzone][junaid]   BRAT I WAS ALSO IN YOUR SITUATION KEEP ON TRYING TILL SOME ONE GET YOU
[DISP-#chatzone][leXie]   brat: hey there. <smile>
[DISP-#chatzone][Poison]   Brat:Having a bad day, but okay otherwise!
[DISP-#chatzone][Brat]   n22: I'm doing just fine. Thanxs. And you?
[DISP-#chatzone][Marie]   Wow Brat, someone wanted to talk !
[DISP-#chatzone][junaid]   KOKOPELLI I AM NOT YELLING IN FACT I AM A COOL GUY

[DISP-#chatzone][n22] good here brat!
[DISP-#chatzone][Brat] junaid: Thanxs. Your kind!!!!!!!!!! =)
[DISP-#chatzone][kokopelli] nobody said you weren't, it's just so loud
[DISP-#chatzone][n22] junaid, some say all capital letters looks like yelling
[DISP-#chatzone][junaid] NO PROBLEM BRAT ANY TIME ALL THE TIME
[DISP-#chatzone][junaid] ALL CAPITAL LETTER MAKES ME PROMINENT
[DISP-#chatzone][kokopelli] well, why isn't your nick in capital letters then?
[DISP-#chatzone][n22] junaid, it makes you weird :-D
[DISP-#chatzone][junaid] DON'T MAKE ME ANGRY OTHERWISE I WILL BLAST
[DISP-#chatzone][n22] junaid, go ahead, make my day
[DISP-#chatzone][kokopelli] ooohh junaid we be scared
[DISP-#chatzone][junaid] N22 YOU ARE THE ONLY ONE WHO TALK TO ME NICELY,
[DISP-#chatzone][n22] hehehe junaid, really?
[DISP-#chatzone][ACTION] n22 is usually pretty violent, but he is burnt out today
[DISP-#chatzone][junaid] OF COURSE, WHO ARE YOU N22
[DISP-#chatzone][Romario] Junaid - kill the caps

Undernetters also make use of the private channel as a strategy for defining situations. This occurs quite often in the instance of behavior which is being defined as deviant, a situation which is discussed in the next chapter. However, the private channel is also used to define situations for newbies who need help; when a newbie appears to be at a loss regarding interaction in the public channel, a regular may message him or her privately to offer assistance. This facilitates the socialization process without disrupting the stream-of-consciousness flow of the public channel. In the example below, Marcia is new to this channel. When she joined, Kaz asked her for her 'stats'. She did not respond, because she did not understand the request. Mollie, a regular on the channel, messaged her privately:

[DISP-mollie][mollie] hi do you know how to play this game? Kes asked for your stats
[DISP-mollie]<Marcia_Br> no. i was just doing the scene from the movie:(

[DISP-mollie][mollie]   this is a game channel. King A (kes) asked for your stats which essentially picks you to play
[DISP-mollie]<Marcia_Br> oh. i'm sorry.
[DISP-mollie][mollie]   no problem...it's a cool channel and cool game....those guys just explained stats...please give yours?
[DISP-mollie]<Marcia_Br> but what's the game?
[DISP-mollie][mollie]   question....we ask each other deep questions is all...my stats for example are 32/f/married/student
[DISP-mollie]<Marcia_Br> o.k., i'll try.
[DISP-mollie][mollie]   you pick someone besides kaz or the bots
[DISP-mollie]<Marcia_Br> o.k.
[DISP-mollie]<Marcia_Br> o.k., but what kind of question do i ask?
[DISP-mollie][mollie]   just pick someone and ask for stats
[DISP-mollie][mollie]   then maybe their stats will inspire a question
[DISP-mollie]<Marcia_Br> o.k.
[DISP-mollie][mollie]   a good q for scout would be to ask her if she'd consider dating someone from IRC and if so whom? Or ask her "Who from IRC would you like to meet in real life and why?"
[DISP-mollie]<Marcia_Br> o.k. will she get mad at me??
[DISP-mollie][mollie]   NO she's cool
[DISP-mollie]<Marcia_Br> o.k.
[DISP-mollie][mollie]   and we all come into this game knowing we could be asked anything

Because this channel is a large one, it is difficult to get to know all of the people. This game is intended to be a facilitator in this respect; it offers a means by which regulars may ask newbies questions in an attempt to assess them. Interaction in a private channel helps a newbie understand and fit in with his or her surroundings. Regulars guide newbies through the maze of social rules in much the same way a parent guides a child in a new social situation.[2]

In addition to using public and private channels as means for socializing newbies, Undernetters also use both types of channels to achieve reciprocation and commitment from other members. As Elisabeth Reid [1991] notes:

"For those who can keep the pace, such stream-of-consciousness communication encourages a degree of intimacy and emotion that would be unusual between complete strangers in the 'real world'. The IRC community relies on this intimacy, on spur of the moment social overtures made to other users. . .IRC users regard their electronic world with a great deal of seriousness, and generally with a sense of responsibility for their fellows." [p. 16]

During the course of this research, the author asked a number of people about the extent to which they can gain a sense, through IRC interaction, that they 'really' know people and whether or not the relationships that develop on-line are meaningful to them. In every instance, the reactions followed along the lines of these sample responses:

[DISP-#35plus]<Marcia_Br> are the people how you imagined them on irc?
[DISP-#35plus][isis]   hmmm
[DISP-#35plus][dcannabis] marcia: nope, not at all
[DISP-#35plus]<Marcia_Br> how so??
[DISP-#35plus][dcannabis] well, just different than what you have pictured in your mind
[DISP-#35plus]<Marcia_Br> are their personalities the same, though?
[DISP-#35plus][dcannabis] yes
[DISP-#35plus]<Marcia_Br> cool...
[DISP-#35plus]<Marcia_Br> is it different to talk to them on irc after you've met them in the flesh?
[DISP-#35plus][dcannabis]   marcia: not really, feel like you already know them

[DISP-oreo]<Marcia_Br>   so you're a regular here?
[DISP-oreo][oreo]   yeah, too regular
[DISP-oreo]<Marcia_Br>   who do you consider part of the group?
[DISP-oreo][oreo]   gonz0, drizz and jac are the three who I know best
[DISP-oreo][oreo]   at least of the people who are here right now
[DISP-oreo]<Marcia_Br>   do you know them in person, too, or just here.
[DISP-oreo][oreo]   no, just here
[DISP-oreo][oreo]   i haven't met anybody in person yet, but i've only been on about a month
[DISP-oreo]<Marcia_Br>   are they friends just like 'real life' friends?
[DISP-oreo][oreo]   I consider several of these people to be my friends just like normal

[DISP-oreo][oreo]   but you always have to wonder a little, 'cause you never know
[DISP-oreo]<Marcia_Br>   what do you mean?
[DISP-oreo][oreo]   I don't think it applies to people here, but there are a lot of twisted people on the net
[DISP-oreo]<Marcia_Br>   twisted how?
[DISP-oreo][oreo]   twisted as in not friendly, not honest, ...

Here, oreo comments that IRC friends are 'just like normal' friends, though he or she has never met any of them in the traditional sense (seen their physical bodies). Oreo goes on to suggest that, while there are some 'twisted' or dishonest people in IRC, he or she does not believe that those types of people are in channel #newchat. This is a very common theme within IRC. In general, people believe that there are a few bad apples within the network, but that: 1) that 'type' does not exist in the channels which they, themselves, frequent; and 2) they can readily spot those deviants and kick them from the channel in order to maintain channel integrity. Regulars of any given channel claim that their channel is the friendliest and that it is their role to keep it that way--to protect it from the bad apples.

While the discussions within public channels are usually light-hearted, they can and do turn, if only for brief moments, to the 'personal and anguished'. And it is this commitment to the belief in one's regular channel as the most friendly and trustworthy that allows for these episodes. In the example below, every effort is made on the part of the regulars to comfort the individual in need. Within this framework, however, the serious matter is never allowed to continue for long periods, and it is usually the individual in need who attempts to redefine the situation back toward the light-hearted:

[DISP-#41plus][PaulH]   mourning ?
[DISP-#41plus][PaulH]   who'd you kill :)
[DISP-#41plus][Stormy]   Cmm....it does say on a good day...this is NOT a good day :)
[DISP-#41plus][Lainy]   hi Marcia
[DISP-#41plus][Stormy]   my dog
[DISP-#41plus]<Marcia_Br>   hi, Lainy
[DISP-#41plus][Stormy]   I killed my dog
[DISP-#41plus][cmm]   Oh... ;(

[DISP-#41plus][Loon]   you what stormy? killed your dog.. oh no...
[DISP-#41plus][vegieman]   Killed your dog?
[DISP-#41plus][TazDevil]   paulh....Stormy had to put her dog down
[DISP-#41plus][cmm]   I am sorry...
[DISP-#41plus][vegieman]   I am sorry too
[DISP-#41plus][PaulH]   stormy : tsk tsk
[DISP-#41plus][Lainy]   awwww Stormy
[DISP-#41plus][Loon]   I'm sorry to hear that stormy, that's a hard thing to do...
[DISP-#41plus][SasQuatch] :(
[DISP-#41plus][vegieman]   do you feel bad stormy?
[DISP-#41plus][TazDevil]   she is very upset by it (understandably)
[DISP-#41plus][PaulH]   : (
[DISP-#41plus][Loon]   I could cry thinking about it.... I had to do that once...
[DISP-#41plus][PaulH]   stormy: me and my big mouth : O
[DISP-#41plus][vegieman]   Lets have a little ceremony for stormys dog
[DISP-#41plus][TazDevil]   paulh .... exactly
[DISP-#41plus][LadyofFe]   re stormy *huggers*
[DISP-#41plus][SasQuatch] re stormy
[DISP-#41plus][cmm]   I agree. We should have an on-channel ritual for Stormy's dog.
[DISP-#41plus][Stormy]   No please Vegieman...thanks
[DISP-#41plus][Stormy]   no
[DISP-#41plus][vegieman]   sorry stormy
[DISP-#41plus][PaulH]   stormy: so sorry.
[DISP-#41plus][vegieman]   wasn't joking
[DISP-#41plus][cmm]   ok, as you wish, but ritual can mark difficult times.
[DISP-#41plus][Stormy]   s'ok thanks everyone....next topic

In this example, a regular who is in need of some comforting from friends initiates the conversation, and the response from the other channel members is immediate. All other conversation is momentarily dropped, while the channel members attempt to help each other. Any individual who does not properly conform to the necessities of that particular interaction is quickly reprimanded by another individual. In this case, TazDevil reprimands PaulH for his flip comment, 'who'd you kill'. The temporarily deviant individual was attempting to utilize the standard interaction formats of IRC in instances which required a different type of interaction; one which would reciprocate the trust placed in them.

While public channels are occasionally used for this more personal style of interaction, such conversations are generally reserved for private channels. Private channels, because they are by definition more intimate, allow for and demand a degree of personal disclosure that is unacceptable in public channels. They are the means by which Undernetters get to know each other as individuals; this type of interaction supports both the light-hearted and serious discussions of public channels. To put it simply, in private channel discussions, one learns who one can joke with and in what way in public, and which topics one should take seriously when brought up by another individual in public.

Fairly intimate self-disclosure occurs frequently during the first private channel interaction with an individual, and for the most part, the participants ask and respond easily to questions that, if put to them on such short notice in 'real life', could be taken as offensive. In an initial private channel interaction, one might be asked his or her age, marital status, plans for a family, and so forth. Or, channel members might confide personal problems or issues with which they are dealing. Within the culture of IRC, these types of interactions can only be considered natural.

It is through such interactions on both public and private channels that regulars and newbies: 1) negotiate standards of interaction (socialization); 2) present selves; 3) form significant and generalized others; and 4) achieve reciprocation and commitment. The ways in which Undernetters utilize the public channel format for interaction necessitates the use of the private channel format, and vice versa. That is, people have constructed public channels as the Third Place; a place to hang out with others and generally just enjoy themselves. In order to accomplish this, however, participants have developed the private channel as a means of establishing enough of an identity for the other that 'just hanging out' is plausible. As Oldenburg [1989] has noted, hanging out is not something that is accomplished with strangers; it requires a knowledge of the other which is extensive enough to allow for feelings of friendship.

The two other elements of Undernet life critical to the establishment and maintenance of this atmosphere of the Third Place are the nick and the accompanying "/commands". The nick and related /commands basically serve as early-warning detectors for possible threats to the friendly format of public channels in two ways: 1) one can stereotype potential 'deviants' with a relatively high degree of accuracy; and 2) one can verify whether the particular nick in question is the 'real' person.

Regarding stereotyping based upon the nick, recall the example from Chapter Four in which REALHUNK enters channel #35plus:

[DISP-#35plus][SERVER]   REALHUNK!~jlaksdl@204.7.222.140 has joined this channel
[DISP-#35plus][LComeno]   bad nick
[DISP-#35plus]<Marcia_Br> lol
[DISP-#35plus][Elinore]   heheee!!
[DISP-#35plus][SERVER]   REALHUNK has left this channel
[DISP-#35plus]<Marcia_Br> gee, and just when i was going to go for it. hehehe
[DISP-#35plus][ProTalk]   Ladies you missed REALHUNK ...
[DISP-#35plus][LComeno]   why do they think a nick like that will help them
[DISP-#35plus][ACTION]   Elinore sooooo upset...missed realhunk
[DISP-#35plus][jeff2]   that used to be my name...
[DISP-#35plus]<Marcia_Br> lcomeno: who knows....
[DISP-#35plus][ACTION]   LComeno was getting ready to kick realhunk

Had REALHUNK not left the channel of his own accord, he likely would have been kicked by LComeno. Nicks such as these are almost always seen as potential threats to the easy-going conversation of the channel; harassment in the form of unwanted messages, either in the public channel or in a private one, is defined as usually coming from nicks such as this who are perceived as 'out for netsex'. However, channel regulars are not completely unforgiving of poor or potentially threatening nick choices; much depends on the definition of the situation as negotiated by the 'offending' individual:

[DISP-#25plus][SERVER]   wenis!~wirtes@204.179.112.220 has joined this channel
[DISP-#25plus][Stevie_B]   hey wenis
[DISP-#25plus][qtpie]   hi wenis
[DISP-#25plus][Therapist] heehee
[DISP-#25plus][wenis]   hello all...

[DISP-#25plus][DarkSpher] Wenis .. unfortunate nick.
[DISP-#25plus][wenis]   Hello steve_b, qtpie
[DISP-#25plus][blondee]   hi wenis
[DISP-#25plus]<Marcia_Br> hehe, dark
[DISP-#25plus][wenis]   hello blondee, what up?
[DISP-#25plus][blondee]   not much!
[DISP-#25plus][DarkSpher] Not the wenis anyway.
[DISP-#25plus]<Marcia_Br> hehe
[DISP-#25plus][blondee]   LOL Dark
[DISP-#25plus][wenis]   25plus, is this the "mature" area?
[DISP-#25plus][Stevie_B]   wenis : hardly :^)
[DISP-#25plus][blondee]   wenis, depends on what you consider mature!
[DISP-#25plus][Cubbi]   hehehehhe
[DISP-#25plus][DarkSpher] No ... just silly.
[DISP-#25plus][DarkSpher] Puberty is a prereq.
[DISP-#25plus][wenis]   At least you guys didn't say, "Hey, that sounds like PENIS!"
[DISP-#25plus][qtpie]   wenis: well we thought about it :)
[DISP-#25plus][ACTION]   wenis thinks about changing his nick
[DISP-#25plus][blondee]   wenis, don't change it, it's great!
[DISP-#25plus][DarkSpher] ... and that's an order!

While REALHUNK said nothing in the channel, thus defining himself in the minds of channel regulars as a 'lurker' and a potential threat, wenis attempts to join in the conversation despite the fact that his nick is commented upon unfavorably. He joins the banter in an acceptable way by joking about his own nick and suggesting that maybe the channel is not as 'mature' as it should be, given the title, #25plus. In turn, the channel regulars also joke about the maturity level of their own channel. As the negotiation of the situation continues, wenis himself suggests that maybe he should change his nick. At this point, the regulars inform him that he need not change it; they accept him and his nick.

Of course, there are channels in which verification of 'true' identity beyond the nick is considered vital for the maintenance of order and friendliness. For example, the #lesbian channel is intended to be for women only, as a means of discussing 'women's issues' and avoiding unwanted interactions with men in the Undernet. While some males, who have proven to

Socialization and Personal Identity

the regulars that they are not a threat to the channel, are allowed in, most are scrutinized immediately upon entry with the /whois and /finger commands. In this first example, moth has just kicked and banned an intruder who was harassing the newbies to the channel, and is offering advice to the newbies about the things to look for in order to know if someone is 'really' a male:

[DISP-#lesbian][Cyllia]   moth: you kicked him off?? powerful command indeed!!!!
[DISP-#lesbian][moth]   i don't have much patience for hng's
[DISP-#lesbian][moth]   so, be very very wary of anyone propositioning you
[DISP-#lesbian][moth]   you can start by doing /whois <nick>
[DISP-#lesbian][moth]   to see if there is a male name there at all
[DISP-#lesbian][Suzie]   you call always tell after you've talked to them long enough
[DISP-#lesbian]<Marcia_Br>   how, Suzie?
[DISP-#lesbian][moth]   for example, do /whois lupus
[DISP-#lesbian][moth]   right suzie--the men are often pretty obvious
[DISP-#lesbian][moth]   well, if you can do /finger <nick>, that can sometimes give you more info
[DISP-#lesbian][moth]   most men will give up quickly or get defensive or try to avoid answering if you really pursue their gender
[DISP-#lesbian][moth]   those that seem very interested in netsexing with you are very likely to be male as well

In this next example, still in the channel #lesbian, an interesting negotiation of sexual identity takes place between Calvin and the channel regulars. As was the case with 'wenis' and poor nick selection, identification through the nick and even the /whois and /finger commands may often be overridden by the individual's ability to redefine the situation in his or her favor:

[DISP-#lesbian][SERVER]   Calvin!~ccartaya@chuma.cas.usf.edu has joined this channel
[DISP-#lesbian][Cass]   Calvin: Women only, please leave
[DISP-#lesbian][Calvin]   I am a woman!
[DISP-#lesbian][Calvin]   I just like Calvin and Hobbes a lot
[DISP-#lesbian][ACTION]   wetness looks at calvin with some doubt

[DISP-#lesbian][ACTION]   Jeannette is checking...
[DISP-#lesbian][Jeannette]   Yes indeed, Carlos.
[DISP-#lesbian]<Marcia_Br>   carlos, calvin
[DISP-#lesbian][Cass]   who is carlos then
[DISP-#lesbian]<Marcia_Br>   that's calvins real name
[DISP-#lesbian][Jeannette-KICK]   has kicked Calvin from #lesbian ^ sorry charlie. ^
[DISP-#lesbian][Jeannette]   hey all, size 11 1/2 C shoe for 190 lb girl sound about right?
[DISP-#lesbian]<Marcia_Br>   yeah, jeannette
[DISP-#lesbian][Jeannette]   11 1/2 is big, but 190 is big too
[DISP-#lesbian][Jeannette]   Calvin checks out I think
[DISP-#lesbian][Jeannette]   she uses some herbal tea called carmencita for pms and cramps
[DISP-#lesbian]<Marcia_Br>   jeannette: what's that?
[DISP-#lesbian][Jeannette]   Calvin, who was here a moment ago
[DISP-#lesbian][Jeannette]   Well, we can grill her some more...
[DISP-#lesbian]<Marcia_Br>   hehe, jeannette
[DISP-#lesbian][Kitiara]   who are we grilling?
[DISP-#lesbian]<Marcia_Br>   carlos is coming back
[DISP-#lesbian][ACTION]   Kitiara takes out her bbq hat and apron
[DISP-#lesbian][SERVER]   Calvin!~ccartaya@chuma.cas.usf.edu   has joined this channel
[DISP-#lesbian]<Marcia_Br>   lol
[DISP-#lesbian][Jeannette]   at 190 she is a BIG girl, so 11 1/2 C shoes is about right
[DISP-#lesbian][Calvin]   hello!
[DISP-#lesbian][Jeannette]   Men width C is *narrow*.
[DISP-#lesbian][SERVER]   Calvin's nickname is now Cristina
[DISP-#lesbian][ACTION]   Cass gets the gasoline and matches
[DISP-#lesbian][Kitiara]   11 1/2 shoe?
[DISP-#lesbian][Jeannette]   for a 190 lb girl.... I'm out of my league here
[DISP-#lesbian][Cristina]   I'm really tall- 6' 3'

    Use of the nick, Calvin, is the first strike against this individual because it indicates a male identity. When Jeannette performs the "/finger" on Calvin and the 'true name' of Carlos is revealed, she is certain that Calvin is a man and kicks him from the channel. Calvin then messages Jeannette in private to plead his or her case, and offers what Jeannette claims is enough evidence of female-ness to warrant allowing him or her back in the channel. When he or she re-enters the channel, Calvin immediately changes nicks to Cristina in order to present a more favorable identity. As the channel regulars continue to question Cristina, they agree that the individual is, in fact, a 'she' and allow 'her' to remain in the channel. For purposes of interaction in this Undernet channel, then, Cristina is 'really' a woman.

While the atmosphere of the Undernet is that of the Third Place, as the examples above point out, and as Reid [1991] has argued,

"IRC is not a 'game' in any light-hearted sense--it can inspire deep feelings of guilt and responsibility. . . .Experimentation ceases to be acceptable when it threatens the delicate balance of trust that holds IRC together. The uniqueness of nicknames, their consistent use, and respect for-- and expectation of--their integrity, is crucial to the development of these online communities. . .users of IRC who flout the conventions of the medium are ostracized, banished from the community." [p. 18]

In the following example, Emily2, who is a long-time regular in the channel #chatzone, has her nick stolen by another individual. When one logs on to IRC, the computer will attempt to connect with the standard nick. If someone else is using that nick, one's computer yields the message, "nick currently in use; select another". This is the first indication that one's nick has been jacked. Emily2, unable to use *her* nick, uses the nick 'zoopy' to gain access to the channel. Upon entering, zoopy sees that someone is indeed imitating her nick, with the intent to tarnish her name:

[DISP-#chatzone][SERVER]   zoopy!~ta0gxr1@corn.cso.niu.edu has joined this channel
[DISP-#chatzone][ChatServ-MODE] Has changed zoopy's mode to +o
[DISP-#chatzone][Emily2]   SHUT UP FAGS
[DISP-#chatzone][NIKKI]   Bad hair day Emily 2
[DISP-#chatzone][Sheena]   Emily2:Are you having a bad day?
[DISP-#chatzone][otis]   is emily2 = gmor
[DISP-#chatzone][Emily2]   STOP TALKING TO ME LOSERS
[DISP-#chatzone][zoopy-MODE]   Has changed *!*gmoriart@*.chi.il.us's mode to +b

[DISP-#chatzone][zoopy-KICK]   has kicked Emily2 from #chatzone ^ |
Emily |^
[DISP-#chatzone][NIKKI]   then leave Emily 2
[DISP-#chatzone][Marie]   Emily2 I like u :)
[DISP-#chatzone][Brat]   Nikki: Funny!! You tell Emily2!!!!!!!
[DISP-#chatzone][stone]   yess she is gone,, hallahuha
[DISP-#chatzone][NIKKI]   thanx n22
[DISP-#chatzone][SERVER]   zoopy's nickname is now EMily2
[DISP-#chatzone][ACTION]   Chryseis wonders if that was indeed the REAL Emily@...?
[DISP-#chatzone][hunter]   hey emily2: what's your problem??
[DISP-#chatzone][Mozart]   Em2!!!! the real one!

In this instance, the channel bot, ChatServ, plays an important role in identifying the first Emily2 as a 'fake'. Channel bots op people based upon server information, not the nick. Thus, when the fake Emily2 entered the channel, she was not given ops, and when the real Emily2 entered under the nick, zoopy, she was given ops by the bot. This flaw in the fake Emily2's identity, combined with the display of personality traits not associated with her by those who know her, lead some of the channel regulars to suspect that Emily2 was a fake.

Bots, public and private channels, ops, nicks and /commands all play important roles in the maintenance of social order and identity throughout the Undernet. Undernetters pride themselves on a friendly system of fair and efficient communications. The institutions and groups which they have developed serve those goals through allowing for an orderly process of socialization of newbies and a means of ensuring the integrity of individual and channel identities. Through such institutions and groups, Undernetters maintain the beliefs and system of values which are the foundation of the Undernet. Though there is always the potential for disruption or miscommunication, as the discussion above suggests, there is a very high degree of order and stability within and across channels. Channel titles provide a general frame within which interaction is defined. Channel ops and other regulars, through a complex system of interaction, maintain order in those channels, and such communication requires a generally accepted set of norms and symbol systems. Further, the fact that Undernetters go to great lengths to appear openly trustworthy lends additional stability to channel and individual integrity. And there is a universal sense among Undernetters that most of the people there are honest, and that those who are not are easily detected and 'deserve' whatever happens to them as a result. Ops can kick and/or ban deviants, just as members of face-to-face communities may informally and sometimes formally ostracize those found to be deserving of punishment.

THE USENET COMMUNITY

Usenet community members accomplish the same basic goals of socialization, the only difference being the particular institutional strategies they use. The key institutions and groups within Usenet, as discussed in Chapter Four, are FAQs, newsgroups, sysops, moderators, message formats and signature files. In particular, this chapter focuses on three specific points with regard to the process of socialization and the presentation of self within Usenet: the use of FAQs, the delineation of basic 'types' of newsgroups, and the use of message formats and signature files.

The six FAQs outlined in Chapter Four are those which have evolved over time, as Usenet has grown, in an attempt to establish the basic norms of acceptable behavior within the community. But because the atmosphere of each newsgroup is at least slightly different from that of every other, the members of a good proportion of newsgroups have developed their own FAQs which simultaneously reflect the spirit of the community in general and meet the particular needs of individual groups. Depending on the group, the FAQ may range in topics covered from a focus on social interaction information specific to the group to a focus on answering commonly asked topical questions relevant to the group's mission. In either case, the goal of the FAQ is to improve the efficiency of communications within the newsgroup by minimizing the need to answer the same questions repeatedly.

In Chapter Three, the point was made that it is useful to consider Usenet through the analogy of rooms. While the room analogy is appropriate, it would not be an over-generalization to suggest (based upon this research) that there are two broad types of rooms, or newsgroups, within Usenet; those oriented toward debate (or in some cases, argument), and those oriented toward problem -sharing and -solving. However, a rigid line between these two types of newsgroups cannot be easily drawn; those one might classify as debate-oriented can, on occasion, have a fair amount of problem-sharing running through the various newsgroup threads. Likewise, those oriented toward problem sharing and solving can foster heated debates on particular subjects. Think of each newsgroup as lying somewhere along the problem - debate continuum.

The alt.support.depression newsgroup is one which lies toward the end of the scale as a problem-sharing/solving group. Many people who are regulars in this group, as the name suggests, are seeking advice and support from others who share a common problem. Because the newsgroup is designed to gather together individuals in various stages of fragility in their personal and professional lives, it is of paramount importance to ward off inappropriate behaviors before they occur. Simultaneously, however, members must be encouraged to ask questions and seek solutions to their own problems; this is the mission of the group. Thus, the following FAQ has been designed by the regulars in an attempt to achieve both of those goals:

Archive-name: alt-support-depression/faq/part1
Posting-Frequency: bi-weekly
Last-modified: 1994/08/07

alt.support.depression FAQ
==========================

Introduction

Alt.support.depression is a newsgroup for people who suffer from all forms of depression as well as others who may want to learn more about these disorders. Much the information shared in this newsgroup comes from posters' experience as well as contributions by professionals in many fields. The thoughts expressed here are for the benefit of the readers of this group. Please be considerate in the way you use the information from this group, keeping in mind the stigma of depression still experienced in society today. . .

alt.support.depression rules for newbies:

>1 There are no Rules.

>2 If you need to ask, you are a newbie. It just means a newcomer to the >
Net or a.s.d (this group). The Net is full of strange and often far from >
wonderful terms. If in doubt, ask, or just use english (see Rule 16).
>3 Netiquette applies, but if in doubt, ask. This is a support group, and >
people will try to help. If you have a real problem (as opposed to a Net >
or group related one), ask and damn the netiquette!
>4 Feel free to loiter round the edges for as long as you like. With luck, >
you will get the answers and help you need without asking. There are >
FAQs posted at intervals (FAQ=frequently asked questions). They >
contain answers to common questions (hence the name). The main one >
is posted at fortnightly intervals; if it's not in the group at present, post a >
request for it. The mini-FAQ should be posted daily. Both are available
>5 If you do ask, say if you really NEED help. There is no rotation of >
helpers who take it in turns to deal with posts. If you have an urgent >
problem, say so. You are unlikely to be ignored anyway, but if you make
it clear you are having a major crisis, you will be up to your ears (eyes?) >
in responses. Use the _header_ or _subject_ and say URGENT or >
HELP or some such. Make us notice.
>6 Don't be afraid to ask for help or information. This is a friendly, >
supportive group. We are here to help each other.
>7 If asking for help, try not to invest too much emotional capital in it. >
While you are more likely to come up trumps, you may just get trumped.
If you don't get a response after a while post again, WITH EVEN MORE
CAPITALS! We _all_ have off days. It is also worth-while considering >
that this is cyberspace. You may have posted, but we may not have >
received!
>8 If you aren't looking for help, but want a place to let off steam, label it >
RANT, and say whatever you like.
>9 If you know what you're going to say before you start, please put in an >
indication as to what you're about to say: parents, lovers, profanity, >
poems etc. It isn't essential, but it can help those with specific problems >
to avoid being hurt. Particularly important ones are: CUTTING; lots of >
people don't like to read about this; HUMOR; another one some people >
like, others don't. You are likely to generate "me too" follow-ups; you
may have a problem none of us has shared, but don't count on it!
>10 Please remember that there are those here who have been hurt by >
religion, and those here who have been helped by it. Please try to be

considerate of the feelings of each. If you need to post in this territory, > put RELIGION in the subject.
>11 If you need anonymity, both FAQs provide details. (Or follow the > instructions at the bottom of an anonymous posting.)
>12 This is a no-flame zone. Try not to flame even the most obvious trolls. > It's usually worth asking why they did it. Even if you don't want to help > them, trolls can have problems like ours.
>13 This is a no-flame zone, but other groups are not! Be careful when > responding to posts which are made to more than a.s.d (check the > headers). If they are cross-posted (posted to more than one group) > please reply only to this group (edit out the others). If you don't, you risk us getting strange and sometimes very unpleasant replies.
>14 Allow for the fact that the Net is world-wide, though most of its users are still in N America. This is an english language group, and has people > reading from it and writing to it from all over the world, not all with > English as a first language.
>15 It _is_ world wide, so bear that in mind when tempted to be parochial. > Sometimes it is the right thing, when the problem is local, but don't > expect your answer to apply across the world. It probably won't.
>16 There is no standard, world-wide english. Don't quibble about spelling or grammar. (If in doubt, get an english-english dictionary. They do exist!)
>17 If giving advice, don't be too dogmatic. Ardent or missionary zeal is > unlikely to go down well, no matter what it is applied to. We are all > different, we all react differently to the same drugs, therapies and > situations, let alone different ones.
>18 Occasionally, members of the group will try to redefine the world. These attempts are, at worst, harmless though irritating. At best they can throw > up much enlightening information and different views of what you may > not even have seen as a problem. Treat accordingly.

For newsgroups which are further along the continuum toward debate, but which may still be generally categorized as oriented toward problem-sharing/solving, the FAQ is often not as explicit regarding which types of interactions are socially acceptable and which are not; group members are capable of tolerating a higher degree of irritation caused by deviant behavior. For example, the following is the (abbreviated) general FAQ for the newsgroup rec.pets.dogs.*. The basic goal of the FAQ is to orient newbies to the basic history and goals of the group. While specific norms of behavior are not stated, it is noted that certain threads, or topics, are highly likely to result in flame wars if and when they are raised. It is suggested that, rather than wasting time with such interactions, newbies read the FAQs on those specific topics and gather the facts before starting a debate:

INTRODUCTION TO REC.PETS.DOGS.*

Table of Contents

* History of rec.pets.dogs.*
* Current Groups and Charters*
* Acknowledgements*
* History of the FAQ's

History of rec.pets.dogs.*

Prior to the summer of 1991, there was a single newsgroup for discussing issues of interest to pet owners. This was rec.pets, and this group still exists for those pets that don't have their own newsgroups.

Rec.pets.dogs (and rec.pets.cats) was formed in the summer of 1991. Joe DiBenedetto proposed the split and collected the votes, which proved more than enough for its official creation.

In the late summer of 1994, another discussion on splitting rec.pets.dogs itself started up, due to the hundreds of daily messages on the group. This split was proposed by Cindy Tittle Moore, and Ronald Dippold collected the overwhelmingly positive vote. The new splinter groups were created on November 9th, and rec.pets.dogs was removed two months later.

ORIGINAL CHARTER FOR RPD

Rec.pets.dogs is a newsgroup devoted to domestic canine issues. This group can be friendly and helpful. Flamewars are generally limited to several topics: crating dogs, training by the Koheler method, docking and cropping, animal rights activism, wolf hybrids and pit-bull (or other breed) dog bans. New readers are advised against starting these topics up on the newsgroup as long, protracted, and inflammatory discussions often ensue. The facts pertaining to these controversial topics are covered in the FAQ's. This is not to say that these topics are forbidden from discussion on rec.pets.dogs, but that regular readers would greatly appreciate it if you checked out what the FAQ had to say on them to make sure you have something new to contribute.

As these two sample newsgroup-specific FAQs suggest, newsgroup members construct FAQs that simultaneously reflect the values and norms of the Usenet community as a whole while attempting to meet the particular needs of individual groups. These FAQs are posted periodically within the newsgroups in order to ensure ease of access for newbies. This is an important aspect of socialization and social control because newbies or those who disregard the suggestions made in FAQs can be admonished to 'read the FAQ', and there is little room, on the part of the newbie or deviant, for creating a legitimate reason for not having done so.

Besides the FAQ, interaction within newsgroups themselves is the key strategy for accomplishing socialization and personal identity. In many ways, though the atmosphere of the Undernet and Usenet differ significantly, the strategies of interaction within Usenet newsgroups and Undernet channels are similar. In any given newsgroup, multiple conversations are being carried on simultaneously, and group members may choose to participate in one or many. Group members download new messages and sift through them using a newsreader. Individual messages appear in varying levels of embeddedness, as in an outline format, and the reader can determine from this format which messages belong to which topic "threads".

While these threads often flow along without incident, this is not always the case. Every newsgroup, regardless of its place on the problem-debate continuum, contains threads which are marked by strong debate. In instances such as those, in which group members challenge the thoughts of one another on the topic at hand, there is always potential for the thread to disintegrate into a flame war. However, flaming and flame wars are a relatively rare phenomenon in Usenet, particularly considering the volume of interaction. This is due in large part to the measures that group members themselves take to work together in defining the conversation as one in which opinions are not really that far apart; Usenetters consistently choose to abide by the norms of their community, as the following example illustrates:

Cinnamon Minx <whitewlf@tiac.net> wrote:
>tshepar@hubcap.clemson.edu (The Hosehead) wrote:
>>Interesting that you think animal abuse and child abuse warrant the same
>>punishment. Are children no better than dogs? Or are dogs just as important
>>as children? Methinks you ought to rethink what you value the most, since
>>there is no rational argument to justify placing animals and human children
>>on the same level of importance.

>I don't think that's what Martha meant...probably more like....pain and
>suffering of any living creature by another's hands should not and will not be tolerated.
>I used to live in a rural area where it was considered fun by many to shoot
>any animals they happened to run across. Not much was ever done about it,
>because it wasn't considered a crime by the police, and was never seriously
>investigated. (Mostly they shrugged their shoulders and gave the "kids will
>be kids" line) I also had a beautiful black lab who was hit by a car
>intentionally in our driveway They had to go about 1/4 mile down it to hit
>him, then they just left him there. He crawled the rest of the way to my house
>with his back broken in 3 places. That kind of cruelty needs to be stopped,
>and if stiff jail time is the only thing that sends the message, so be it. BTW, we never found >out who did it.

>Cinnamon Minx

Dear Cinnamon, thanks for understanding what I really meant. Please read my reply to hosehead. I think his name is very fitting.

Martha

tshepar@hubcap.clemson.edu (The Hosehead) wrote:
>Martha Swartz <maswartz@earthlink.net> writes:
>>I just saw on the ABC evening news that a man in Florida got 15 years (had
>>a prior felony conviction) for beating and essentially killing a 2-month-old
>>lab puppy.

>>I have to say I think this is a step in the right direction. Now, if we can only
>>get the same kind of sentences for all animal *and* child abuse cases.
>>Feel free to cross-post this to the dogs group. This is just where I usually
>>post.

>>Martha

>Interesting that you think animal abuse and child abuse warrant the same
>punishment. Are children no better than dogs? Or are dogs just as important
>as children? Me thinks you ought to rethink what you value the most, since
>there is no rational argument to justify placing animals and human children
>on the same level of importance.

Of course children are more important. My point was that it's rare that anyone gets 15 years for animal OR child abuse. I feel that anyone who abuses a helpless creature, two legged or four, should get maximum time allowable by law, but check the laws in your state, I'll bet you'll find that the Department of Child Services constantly attempts to "keep families together, thus placing the same kids in the same danger over and over again, instead of just throwing the abuser in jail where they should have been from the very beginning.

Martha

In the first two posts in this thread, Martha and Cinnamon conclude that Hosehead has misinterpreted Martha's point regarding animal abuse and child abuse. And even though Martha, in particular, is frustrated with Hosehead's comment (implying how she feels about him with a comment about the appropriateness of his nickname), both she and Cinnamon attempt to re-define the situation within the bounds of Usenet norms. They do not respond with flames. In other posts within this thread, other newsgroup regulars (Kate and Cici) continue the discussion in defense of Martha and Cinnamon, while taking it a step further (in terms of equating humans and other animals) thus establishing a situation in which everyone (except Hosehead) is in relative agreement:

In article <8042-196271908@inferno.com>, shamrock@inferno.com writes:

>I do think that both animal and child abuse warrant the same punishment
>because the victims are both defenseless creatures. If the authorities were to
>take animal abuse in children, young adults and adults more seriously it
>would circumvent it escalating to abuse to humans.
>Abuse is abuse it's wrong.
>Kate

Agreed. I also recently read the results of a study conducted over quite some number of years (sorry, can't cite the source at the moment, but maybe someone else saw the same article) which studied serial killers, mass murderers, whatever you call them, with particular emphasis on those who not only killed but tortured their victims. I was particularly struck by the mention that well over 95% of these killers said that they STARTED OUT by torturing and killing animals -- one said that he would answer "free kitten" ads in newspapers. (time out to go throw up) If jailing those who torture and/or kill animals for their own sick pleasure will save one human life (let alone the animal lives), I'm all for it.

And please -- don't try to "equate" pets with children. It's not even apples and oranges, it's apples and bicycles, you cannot compare or 'rank' them in any rational fashion. I've both children and pets. I love my children, I love my pets. It's not the same love, any more than I love my children the way I love my husband or my siblings. It's impossible, so please don't waste your own precious energy (not to mention bandwidth) attempting the comparison.

Cici[3]

Rec.pets.cats is typical of other newsgroups in the sense that each newsgroup has certain 'themes'; the threads within each group tend to revolve, in general, around a standard set of issues. People are encouraged to bring new insight into these themes, and to begin new threads within them. However, the overall discussions change very little over time. That is, one could choose not to log in to rec.pets.cats (or any other newsgroup) for a long period of time, and then re-enter only to find that, while the details may have changed, the common themes remain. One definitely gets the sense, even after time has passed, that one is going back to the same 'place'. There is a high degree of continuity of people and topics.

Another way in which Usenet newsgroup strategies are similar to those of Undernet channels is the way in which newbieness is dealt with in the context of defining the situation. Undernetters make important distinctions between 'true' newbies and those who are knowingly violating channel norms. The same is true with Usenetters; it is typical for newbies to openly acknowledge their status in the hopes of heading off any potential problems they may cause by posting to a new group (and this is particularly true in the case of newsgroups that are oriented toward problem -sharing and -solving). In the example that follows, an individual who is new to rec.pets.cats has questions, and is in urgent need of answers. The individual begins the post with an open admission of newbieness as a strategy for defining the situation in which she finds herself:

In article <4809nc$1hq@newsbf02.news.aol.com>, juliannaf@aol.com (JuliannaF) writes:

>>I sure need some help and this is my first Usenet posting. I have a cat that
>>is chewing through all electrical cords he can get his teeth on, including
>>power cords. I have tried "Cat No", and tabasco sauce, all to no avail. The
>>best I can come up with is wrapping all my cords with aluminum foil, and
>>now I'm afraid I will short out my hard drive with static electricity. I
>>probably should post this to a "Dog group", as I have never run into this
>>problem (he even ruined a pair of shoes!!). Does anyone out there have any
>>suggestions? Christmas is coming.

Try a swimming pool supply house. You can buy vacuum-hose by the yard from these folks. Split the hose, put the cords inside the hose, then tape the hose back together (duct tape holds the world together, you know) I'm not saying your cat can't chew through a vacuum hose, but it will take a while, and I'm sure you'd rather check for teeth marks than for sparks!

Cici and the Furry Texans

Compare this scenario to one in which an individual who is fairly new to the newsgroup talk.origins, Justin, attempts to respond to the post of a regular, skeptic, regarding the key thread in the newsgroup, creationism versus evolution. In this instance, Justin makes clear through his responses to skeptic that he is new to the group (by displaying a lack of understanding of points made in the group's many FAQs) and an obvious regular to the group (though a name is not signed) attempts to help Justin by defining for him the general context within which such discussions take place:

In article <48e3o7$e9e@lynx.unm.edu>, Justin Hardin <jmhardin@unm.edu> wrote:

>In article <482hsd$9e0@ionews.ionet.net>, skeptic@ionet.net says...

>> A few months ago I started the thread "Creationist Folly". In the original
>>posting I asserted that creationist where merely those who where ignorant
>>of biology. I note that not a single creationist even claimed to have a
>>biological education, they merely argued from their lack of biological
>>knowledge. Do any of you creationist have a sound biological education. I
>>would enjoy debating these issues but I am frankly tired of trying to give
>>creationists a rudimentary biological education so that we can debate.

>Well, I'm not a creationist...exactly...but I believe it's possible. I've studied
>biology all the way into my second year of college, have a sound background
>in the fundamentals of evolution, and guess that you are both patronizing and
>condescending to anyone that doesn't believe in your personal views.
>Creation was the big bang. Evolution was everything after.

Then you are probably *not* the kind of Creationist that skeptic@ionet.net was referring to, and who regularly get flamed crispy on talk.origins. The minimal creationism that said "God started and sustains the Universe" is a purely philosophical position, and not controversial (on t.o, at least; a.a is a different story ;-). At any rate, science has nothing to say about it, one way or another.

>Why are Creation and Evolution mutually exclusive? Answer: THEY ARE
>NOT! Please, try and respect the views of others...to disagree is one thing, to
>patronize and condescend is another!
What talk.origins mostly deals with is Young-Earthers, Flood Geologists, Theistic Anti-Evolutionists and such who propose wholesale revisions of Natural History, claim evolution could not have happened, and make other such "scientific" claims. Almost always, they have most of their facts wrong, or their logic is hopeless. Different views should usually be respected -- but many of these people show minimal respect themselves, accusing their opponents of stupidity and fraud and condemning them to hell-fire. Arrogance in support of demonstrably wrong positions deserves what it gets.

Even though it is clear to the regulars in this group that Justin is a newbie who has displayed lack of knowledge about the newsgroup and its context, the response he receives is courteous. He does not acknowledge newbieness, but he also does not openly or aggressively challenge the posts of the regulars; his comments question Skeptic's statements, but with the appropriate degree of politeness.

Yet another interesting similarity between Undernet channels and Usenet newsgroups is that newsgroups oriented toward problem -sharing/solving often function in the same ways as private channels function within the Undernet. That is, while debate oriented groups are similar to public channels in that they greatly limit or in some cases exclude discussion of personal issues, Usenetters make use of various threads within problem-oriented groups to achieve a very high degree of self-disclosure and reciprocation from others. In the following (abbreviated) example, an individual with lurker status in the alt.support.depression newsgroup decides to post:

I have been lurking here awhile, and it has been a major factor in my recovery to know there are others that feel the way I do. Thank you all, so much. I have emailed some replies when I was *up* but I'm feeling very blue today and need an outlet - this newsgroup is a good one because I won't feel like I'm being a burden on anyone - you can delete me and I will never know! (this is a semi-joke)

I have started therapy and am on Prozac for depression and Xanax to control anxiety attacks and a touch of OCD (mainly cleaning, straightening pictures, setting clocks. I have on more than one occasion taken the washing machine, the dishwasher, the dryer, and the garbage disposal apart to make sure they were absolutely clean. Once I hung striped wallpaper in the bathroom and when it turned out crooked at the corners I went into a shrieking fit and ripped it all down. I've learned not to ever try striped wallpaper again!) I am getting a little better, I think. I don't cry as often, and am a lot more patient with my little girl. I have (through therapy) traced a good bit of my depression to very low self-esteem. I am smart and not bad-looking, but have gained about 30 pounds since I had my child, and at 165 feel pretty bad about myself. I can't seem to get motivated or find the energy to exercise, except in small bursts of frenzy which only last a day or two. Gourmet cooking is a hobby, and it is really hard to lay off the butter and cream.

My job is also taking its toll. When I started, I was really happy to find something in my area (a relatively small town - too far to commute to _____) where I could do the work I wanted to do (Windows programming). Since I joined the company (a pretty large insurance firm), there have been major shakeups in top management, middle management, and budget. The team I work on (all men) are supposed to be working together to explore new technology, but they have all been here together for some time, and I feel very, very excluded from the group. I am so tired of begging and begging for something to do. They treat me like a little kid - sure honey, you can work on this little puzzle - while they're all working together on new stuff.

Anyway, if anyone reads this, I would love some email from people with similar problems. Why do I sink so low when I have so much going for me? I wrestle with this one a lot in therapy. I am somewhat lonely; we just returned to this area after being away for three years, and all our old pals are gone. We're both so busy that it is hard to make friends, and my husband and I would like to be more sociable. We're not particularly religious, but my therapist thinks church may be a good way to meet people. Don't know about that one.

Keeping the black dog at bay for now - (name)

The following samples were among the responses received:

Hi new gal,
My name is Ame. I read your letter and can empathize with your situation. I have OCD and Depression. I have a supportive husband, parents who try to help and a couple of close friends.

I feel down and get the "Guilts". I beat myself for not being as active as my friends or family. Thoughts are always going through my brain.(the O of OCD.)

If you would like to rant and rave together, feel free to e-mail me.
Support and understanding,
Ame

And:

I'm a new gal too,who would probably do some good(hopefully) by some old-fashioned ranting and raving.

It seems you practically mirror me, I'm new here and am married and have two wonderful children. People ask me why I get depressed when I've got so much going for me. Without sitting them down for two days to explain, I never bother.

I would love to here about how long you've been getting depressed and how it all began? If you want to contact me personally you can get me on (address deleted)

 Clearly, Usenetters often invest a substantial amount of their selves in their newsgroup interactions. This is the case not only with heavily problem-oriented groups, but with groups toward the middle of the problem-debate continuum and even for those that are heavily debate oriented. Consider the following example, in which the regulars of rec.pets.cats discuss the loss of a pet:
Carla Ann Hass <CAH19@psuvm.psu.edu> wrote:

>My sister and her husband's cat, Daisy, died _very_ suddenly on Friday. She
>was just five years old! She had a history of bladder infections and seemed to

>be developing a new one, so Scott took here to the vet Friday morning. Other >than the bladder infection symptoms, she seemed her usual, wonderful self. >Scott had her in a carrier and had to sit in the waiting room for a few minutes. >He heard her cry a bit, but she had been to the vet lots of times and was >always very apprehensive. Then he heard a bump, but thought she was just >trying to get settled a bit. He went in to the examining room about a minute >later. When the vet opened the carrier, Daisy was lying on her side. The vet >took her out, realized she wasn't breathing, and tried to do CPR but it was too >late. She was dead! The vet said that the most likely explanation was >cardiomyopathy, but he did not do an autopsy.

>Thanks for listening. Any experience or insight would be a big help right now.

>Carla (and Tasi and Mena - Daisy's cousins)
>State College, PA

Tell them not to blame themselves! If it *was* cardiomyopathy the heart attack was sudden and total. My friend's black kitty died this way (on her pillow at am!) she came running down the hall to my room yelling "Borya's dead"! and I tried to see if he had choked on something- she tried CPR- nothing helped. She DID have the autopsy done (other cats in house, one is the black ones littermate). What killed him was a massive heart attack. He did have cardiomyopathy (either a thinning or thickening of the heart muscle walls).

This disease is genetic- but not all kittens develop it if a parent has it. I am so sorry about Daisy. Tell them not to feel guilty-it wasn't their fault.

Amber

Another response was:

In article <95316.114708CAH19@psuvm.psu.edu>, Carla Ann Hass <CAH19@psuvm.psu.edu> writes:

>Thanks for listening. Any experience or insight would be a big help right now.

>Carla (and Tasi and Mena - Daisy's cousins)
>State College, PA

I have no answers, but please accept my heartfelt (heartbroken) condolences on the untimely loss of Daisy. I'll send a message to Nuisance over the Rainbow Bridge to watch for her and take care of her.

Cici

And:

My heart goes out to the owners. I lost my little Punkin' in June, not to cardiomyopathy, but cancer. I cry every time I hear of a pet loss. It's devastating. My sympathy to your family.

Lisa Daly -- Southern New Jersey

While the self is invested in problem oriented groups through personal statements, the self is invested through a different strategy in debate oriented groups; the self is invested in statements of opinions and claims to knowledge. Significant others are established not so much through statements of empathy on the part of others, but through either statements of support or challenge to those opinions and knowledge claims.

For example, the talk.origins newsgroup is a newsgroup oriented heavily toward debate. And though issues of a 'personal' nature are never discussed, members' senses of self are still very much at stake. Often, what an outsider may mistake for a 'meaningless' flame war is, in fact, a means of establishing and reciprocating the self in relation to the other. As Baym [1995] notes, ". . .even the seemingly antirelational practice of flaming can be reinterpreted as a kind of sporting relationship. . .flaming might be compared to forms of ritual insults that are popular in children's peer groups and serve to define them as members of that group. . .Flaming occurs even when there is no anonymity, and might in some cases be better understood as a form of relationship to be enjoyed at an emotional distance than as a form of social paralysis. . ." [p. 158]

As the following example illustrates, such interactions are not 'meaningless' flame wars, as might be assumed by those who lack an understanding of the context. Rather, this sort of interaction is, quite simply, the standard means by which these individuals achieve legitimation of their Usenet selves. They start arguments, they take sides, they attempt to refute another's points, and so forth:

Subject: Re: Let's show compassion- ignore Nyikos
Date: 29 Nov 1995 22:57:04 GMT
From: nyikos@math.scarolina.edu (Peter Nyikos)
Organization: University of South Carolina - Columbia - Computer Science
Newsgroups: talk.origins
References: 1 , 2 , 3

Not "compassion"-- mercy. Mercy to LaBonne.

labonnes@csc.albany.edu (S. LaBonne) writes:
>In article <49cflt$2ud@netnews.upenn.edu>,
>Matthew P Wiener <weemba@sagi.wistar.upenn.edu> wrote:
>>In article <48tl2l$t5j@redwood.cs.sc.edu>, nyikos@math (Peter Nyikos)
>>writes:
>>>Wiener makes another false allegation in this thread, about having refuted
>>>what I said about "t=1 IBD" three times already. He has not even refuted
>>>it once.

>>You're lying.

>Technically, it would be true enough to say that Nyikos has never said
>anything on this topic that even _makes sense_, and to that degree his
>effusions might be said to be irrefutable. ;-)

You are both lying. Or rather you both would be, were it not for Steve LaBonne's smiley.

Matt has made a big undocumented smokescreen about how you, Steve, actually were talking about "t -> t+1 IBD", but if he were to actually quote you in context, it would become clear that you SHOULD have talked about that all the time, but what you did was to explicitly use "t=1" in your definition of IBD and then broach a brand new subject, not dealt with by me before, about the

spreading of a single PURELY HYPOTHETICAL "altruism gene" thru a population, to which "t=1 IBD" *is* relevant.

>Peter, wouldn't it involve less expenditure of energy, and far less loss of face, >for you to go away and study an intro genetics textbook before trying to >continue this discussion?

For you, yes.
Or if you insist on
>continuing to post, why not try holding up your end of the discussion with >Michael Robinson?

I am doing that quite nicely. Too bad you are not interested in exploring the truth with me, never were, otherwise you would be able to hold up your end of the discussion there too.

Michael finally replied to the first of two posts of mine yesterday, and his post still has not appeared in my newsreader. If it doesn't show up by Friday, I'll go to work on the e-mail copy he sent me.

He seems to be a remarkably patient person, to
>the extent that he might actually succeed in teaching you something.

Fortunately, we don't have an adversarial relationship, Michael and I, so your, wording is most inappropriate and offensive in describing what our conversation is all about.

But that's what you want your conversation with me to be all about, isn't it? From the very beginning, your whole purpose seemed to be to build yourself and your net.cronies up by tearing me down; and as I exposed one dishonest act after another by you, your main goal seemed to be to avenge yourself on me, by hook or by crook.

You found some very willing accomplices in Matt and Alexey, but even they can only fool some of the people some of the time.

>He is also the only person other than Matthew who is willing to continue
>responding to you- or haven't you noticed that, Matthew's insults aside,
>you've been soloing for quite a while now?

Of course, you are lying. Alexey and a number of others have responded, and Michael hasn't even responded as often as Wade Hines and "killer yeast". And now things seem to be picking up again, with Andrew McRae and Richard
Harter once again getting involved in a conversation I began a while ago. And there are other respondents.

By the way, how many people are responding to you these days? Not that I really try to evaluate people by that kind of yardstick, but you seem to revel in these kinds of superficial Usenet yahoo evaluations of people.

>[...] Anyone who _still_ doesn't grasp the meaning of the statement that a
>child inherits half its genes from its mother and the other half from its father,
>and who

Your "anyone" is a figment of your mendacious imagination.

>thinks that "complementation group" is a "bizarre" definition of a genetic locus,

Flamemeister Vladmeister did not use that term, he talked about "mutations that complement each other". And I did not say "bizarre", I said "peculiar".

Funny how you and Alexey both saw that, and did not follow up to it, and neither did Vladmeister, even though I reposted it for him when he made noises about having missed another post I did about the same time.

Cowardice?

Remaining display of mendacity, insincerity and/or hypocrisy by LaBonne, deleted. LaBonne could probably have become a successful est "facilitator": that kind of pseudo-psychology con game is just the thing he seems to be a master of.

Peter Nyikos   -- standard disclaimer --
Professor, Dept. of Mathematics
University of South Carolina
Columbia, SC 29208

    In what MacKinnon [1995] refers to as the 'cycle of statement and response', selves are established. In order to attribute such interactions to 'reduced social cues', or flaming for the sake of flaming, one would have to ignore the context of the interaction. This sort of interaction is the norm in the talk.origins newsgroup. People who flame for the sake of flaming tend to flame once and move on. They do not become regulars in a group and devote substantial amounts of time and energy to developing their understandings of technical and scientific issues simply to flame. The only way one can account for the persistence of such interactions is to acknowledge that, for the participants, there is a substantial amount of the self involved in those interactions.
    Additionally, with regard to the formation of significant others in Usenet newsgroups, Usenetters, like Undernetters, consider the people they meet through Usenet to be just as significant as those they meet in 'real' life, as the following post demonstrates:

In article <Pine.A32.3.91j.951119131126.87940H100000@homer24.u.-washington.edu>, Stacy Hill <shill2@u.washington.edu> writes:

>I've got a question:

>This girl that I've been talking to on the internet just died. I've never met her
>or even talked to her on the phone, but it still upsets me that she's gone.
>Should I be feeling this way about someone I've never even seen???

I'm sorry that this has happened to you, and I think it is more than ok to feel upset. I would be extremely upset if any of my email buddies died - hey, I get concerned when they catch the flu! She was a real, genuine person, and just because you haven't seen her or heard her voice doesn't make her any less real. Sometimes it's easy to think that the words on the screen are simply words, but I've always seen those words as the thoughts and dreams and everything else that make up people - those words that she spoke to you on the computer will be with you forever; that is something to treasure always, and miss when it is gone.

zephyr...

While Undernetters construct their channels as the Third Place, a place in which they can, for a time, forget about other arenas of their lives, Usenet newsgroups are used, in many instances, to help define those other areas of life. So, beyond utilizing newsgroup associations to establish Usenet identities, members also utilize those associations to establish and reinforce their identities in other areas of life. This is particularly evident in the problem-oriented groups such as alt.support.depression. Below, Donna submits a post in which she questions her practice of 'cutting'. Newsgroup regulars attempt to reassure her through defining her position in the larger society:

I stopped for a few days because it felt numb where I was cutting, but then I cut somewhere else. I can't (or don't want to??) stop - altho I say I do. Still feeling numb OR depressed - I must be afraid to feel. I don't know why I post these messages - pity? sympathy? or just attention-seeking?

Donna

In article <480r4g$cor@ixnews4.ix.netcom.com>, singingd@ix.netcom.com (Donna R,) says:
>I don't know why I post these messages - pity? sympathy? or just attention-seeking?
>Donna

Or maybe here you can "talk" to people who have some understanding of what you're going thru and who will regard you and your suffering without prejudice.

I'm sorry to have to say this, but society still treats us like "loonies". Have we bought into this school of thought?

Somebody posted an article about us "wallowing in self pity" a while back. I say baloney! Because we are going thru this and we need to "let it out" or just talk to somebody we shouldn't have to feel as if we are looking for pity or we are seeking attention...

Bob Dubery
Johannesburg
South Africa

And:

Donna, there is _nothing_ wrong with trying to get your needs met. You're not imposing on anyone by seeking attention. People have the choice in this forum whether to respond to your requests, and since your needs are legitimate (even though society has stigmatized every need except for eating and sleeping these days. . .), keep posting about whatever you want feedback about . . cutting or otherwise.

Some people do use cutting as a means of making their inside pain more tangible so that it can then be taken care of, by others or by themselves. There's nothing wrong with this way of coping. There are other ways of 'cutting,' such as pain-revealing artistic works or kicking the sh** out of a plastic garbage bin so that everybody can hear and see you hurting. I have substituted these kinds of things for some of my self-mutilation because I have a tendency to form really noticeable scars, but as long as I have needs, I will not stop trying to meet them. Don't you, either. It's your right.
****
K a t e . . . . kmlF92@hampshire.edu
Green-eyed girl-jock medical anthropology student with interest in adolescent health, penchant for the color charcoal grey, and irrepressible maternal

instinct. S h e b r a k e s f o r r a i n b o w s
****

The final two strategies for achieving social order and legitimation of the self in the Usenet community are the signature file and newsgroup message formats. Almost all articles which are posted to newsgroups contain, at a minimum, header information regarding the identity of the poster.[4] Additionally, they also contain signature files with supplementary identity-related information. This is a necessary strategy if Usenetters are to maintain a sense of efficiency, order and identity within their newsgroups. Whereas Undernetters chose nicks for themselves and then asked each other questions of a 'getting to know you' nature (consistent with the easy-going atmosphere), Usenetters need access to this information without wasting time asking each other about it. Not only does such information allow posts to be organized into manageable and logical threads, it also allows Usenetters to get a sense that they know one another well enough to carry on newsgroup conversations to begin with.

A number of sample signature files were given in previous chapters (as well as sigs contained in posts throughout this chapter) which are representative of Usenet. Though Usenetters are not required in any technical sense to create sigs, most of them choose to do so. At a minimum, those sigs contain identity-oriented information. Additionally, they may contain quotes that a particular individual finds noteworthy, or in some instances, they may contain figures constructed through use of various keyboard symbols.

One can imagine that there is almost no end of possibilities for creating unique sigs, and those sigs could potentially grow to enormous size, relative to the amount of text in a given post. In order to minimize the likelihood for such creative 'wastes of bandwidth', Usenetters have developed norms regarding the acceptable size of sigs. Those who violate those norms are subject to various forms of social control, ranging from the sarcastic comment to the more serious flame. To quote Dear Emily Postnews (Templeton):
Question:
"Dear Miss Postnews: How long should my signature be? -- verbose@noisy"

Answer:

"Dear Verbose: Please try to make your signature as long as you can.

It's much more important than your article, of course, so try and have more lines of signature than actual text.

Try and include a large graphic made of ASCII characters, plus lots of cute quotes and slogans. People will never tire of reading these pearls of wisdom again and again, and you will soon become personally associated with the joy each reader feels at seeing yet another delightful repeat of your signature.

Be sure as well to include a complete map of Usenet with each signature, to show how anybody can get mail to you from any site in the world. Be sure to include ARPA gateways as well. Also tell people on your own site how to mail you. Give independent addresses for Internet, UUCP, BITNET, ARPAnet and CSNET, even if they're all the same.

Aside from your reply address, include your full name, company and organization. It's just common courtesy--after all, in some newsreaders people have to type an entire keystroke to go back to the top of your article to see this information in the header.

By all means include your phone number and street address in every single article. People are always responding to Usenet articles with phone calls and letters. It would be silly to go to the extra trouble of including this information only in articles that need a response by conventional channels!"

Besides the sig and header information, the information contained in the body of the post is critical to maintaining an orderly dialogue within newsgroups. It is a norm in Usenet that, when one is responding to an article, one should begin by documenting the article ID number to which one is responding, and then select which portions of the post to 'snip' from the reply. That is, when an article is sufficiently long, one should delete all but the most relevant (to the reply) points in an attempt to conserve bandwidth. In the sample texts above, responses generally begin with a statement such as, 'in article (article number) so and so wrote:'. Then the embedded text is marked either by : or > and the response has no markings.

While this is an excellent strategy for improving the efficiency of Usenet, it can cause problems in terms of identity for those involved. In short, if one does not 'snip' properly when responding, a great deal of confusion may result regarding who said what. Consider the following example:

In article <(null)>, marco@nucleus.com (Marco Migotti) wrote:

|In article <48545k$kg1@athena.ulaval.ca>,
|Alex van Chestein <a.vanchestein@gmt.ulaval.ca> wrote:
||
||
||>>Alex van Chestein
||>>Alexandra,
||>>I've been following this thread, waiting for someone to give you a more
||>>useful advice than "dump the boy-friend", but unfortunately, the mentality
||>>displayed by Janette is pervasive. IMHO, a man who is a caring person, in
||>>a medical professions and, most importantly, with whom you can
||>>communicate well will not be so easy to forget/replace. "Stand by your
||>> man" as the song goes, and remember "...and after all he's just a man".
||
|
|After the way this guy is acting it would probably be the best thing for the
|both of them. If he doesn't like the cat and the cat doesn't like him then it
|won't make for a happy relationship. If the guy is this abusive before a
|potential marriage then he will be worse afterwards because he will feel more
|at ease and act more like himself. It isn't just an issue of your boyfriend
|hating your cat, but, it is a deeper issue of lack of respect and consideration of
|your feelings. You said that he is caring... but if he doesn't even respect the
|biggest things of your life then he has shown that he doesn't care for you at
|all but is in it for something else. Don't let anybody tell you that he can't be
|replaced, since he isn't really communicating with you other that he hates
|your cat. Be patient, that hey, maybe you could tell him that we should be
|just friends until we can, if possible work out the problems. Maybe even get |a
|third party to mediate if he is worth it to you... after all you don't want to
|loose him if it can be avoided. But if the both of you don't do anything about
|it, it could develop into something far bigger. Good Luck!
|
|Marco
|

||||>>Alex, i'm really curious about how this turns out, please drop me a line or
||>>post. Thanks, and i hope that this helps. -ilya
||
||
|| Excuse me, but...
||
|| My name is Alexandre van Chestein. I'm a 15 year-old boy, the one who
||posted those "HELP me convince my dad to adopt a cat" posts. The original
||poster's name I don't remember, but I'm not her. This happened when
||someone quoted only my name, not my message, in her reply. So, now that
||this is cleared up, I hope we can continue giving advice to this young woman.

Examine the top of the body of the text, where there are two lines marked with double "ll", but there is no text next to them. This is followed by Alex's signature and immediately after that by the name Alexandra. In this instance, because someone 'snipped' Alex's entire post, except for his sig, he has now been mistaken for an individual named Alexandra and people are attempting to discuss his boyfriend problem with him. Alex is forced to re-post the article in which the mix-up occurred in order to get people to understand what the problem is.

Given the high volume of traffic across Usenet on a daily basis, such mistaken identities due to editing errors happen surprisingly little. When they do, it appears that the individual's involved are quick to point out the error; it is important to do so in order to maintain one's identity in the newsgroup. And because newsgroup postings are available for extended periods of time before they expire, it is easy enough for people to review past postings to determine the point at which an error was made.

Newsgroup-specific FAQs, the newsgroups themselves, signature files and message formats all play important roles in the maintenance of social order/identity throughout Usenet. Usenetters pride themselves on an efficient system of communication through which people can meet with others to discuss similar interests. The institutions and groups which they have developed serve those goals through allowing for an orderly process of socialization of newbies and a means of ensuring the integrity of individual and newsgroup identities. Through such institutions and groups, Usenetters maintain the beliefs and system of values which are the foundation of Usenet.

Though there is always the potential for disruption or miscommunication, as the discussion above suggests, there is a very high degree of order and stability within and across newsgroups. Newsgroup namespaces provide a general frame within which interaction is defined, and the regulars, through a complex system of interaction, work to maintain order in their groups. When the situation warrants, Usenetters engage in and respond to a very high degree of self-disclosure; this sort of investment and legitimation of the self serve to promote the feeling of membership. Usenetters, like Undernetters, are committed to their identities and work to protect the identity and integrity of the group (and other members as well) against those who are defined as outsiders or deviants.

THE FIDONET COMMUNITY

As with IRC and Usenet, each Fidonet institution discussed in Chapter Four--named conference areas, FidoNews, conference rules, BBS sysops, moderators and taglines--serves a purpose in maintaining the orderly functioning of the Fido community. With regard to the processes of socialization and identity formation in day-to-day Fidonet life, however, this chapter focuses on two specific points: 1) Echo conference areas and the role of conference moderators; and 2) log in registration requirements and message formats.

The Fidonet community shares with the Undernet community an atmosphere of relaxation. While individuals are expected to remain relatively on-topic in a given conference area, allowances are made for the fact that there is an ever-increasing influx of newbies into the community. And, like Undernetters (and in contrast to Usenetters), Fidonetters have developed a culture in which newbies are welcomed and taught the basic rules as they go. Thus, rather than developing an elaborate system of community-wide FAQs, Fidonet founder Tom Jennings set down the general guidelines of BBS community life in an early edition of FidoNews:

> "Bulletin board etiquette is really no big deal, and I hope you don't get the impression that I'm trying to make an issue out of nothing. . .

In the dark ages of modems (pre-1982 or so) there were so few bulletin boards and users that there basically wasn't a problem. You somehow managed to get a modem, got a bulletin board number from a friend, and started dialing. You got nervous and made a mess of the message base, and if you were real unlucky, crashed the board. Everyone knew you were "new", and so were tolerant while you learned how to get around. Crashers and trashers weren't really a problem. . .

These days. . .Instead of a trickle of new users, it's a torrential downpour. New users outnumber old timers on many boards, and that fact probably won't change. . .With the good comes some small problems; the previous "hack at it 'till you get it right" attitude doesn't work on today's overloaded boards that might handle 50 or more calls a day.

Unfortunately, it frequently becomes a situation like a traveler to a foreign country who is totally unfamiliar with local customs. Visitors embarrass themselves by saying the wrong thing, or insult the locals with totally inappropriate reactions. . .

In face to face encounters with people that you don't know well, there are thousands of "unwritten rules" that just about everyone follows. . .A big problem with modeming is that you miss all non-verbal communication details; eye motion, facial expressions, and other cues that help convey otherwise difficult or embarrassing information. You have to make up for this in other ways.

About the only hard and fast rule of BBSing is READ READ READ READ!!! . . .Get an idea of what kind of people are typing things, and a feel for how "touchy" the crowd there is. Bulletin boards are no different than a local bar or whatever; a particular crowd develops, you just have to choose where you hang out. . .

Keep in mind that some things that are wonderful person to person can be absolute disasters in print. . . bulletin board messages are like someone speaking in a monotone, with no pauses between words, behind a black curtain, recorded on video tape a week ago, and played on an out of focus black and white TV set with a dirty screen. From across the street. . .Again, bulletin boards are no different than any other group of people, except the lack of fine detail and the time difference." [Jennings, vol. 2-30, pp. 1-3]

The basic point of this discussion of BBS etiquette is that one should be nice, do the best one can technically, and 'give it a whirl'; if each individual user follows those guidelines, things will work out. Being nice and doing the best one can consist, by and large, of staying relatively on topic within conferences and contributing positively to the group, however that is defined in a given situation.

Remember that Fidonetters make a distinction between 'annoying behavior' and 'excessively annoying behavior'. This distinction is made because the community as a whole anticipates that newbies will unwittingly commit annoying behavior, but only under certain circumstances will this behavior warrant the definition of 'excessively annoying', thus requiring harsher forms of social control to be utilized.

All newbies, regardless of position within the community, are expected to make mistakes and are therefore granted a certain amount of tolerance as they learn. Consider the following example of newbieness, in which an individual (Matt O'Shields) in the Star Trek: The Next Generation conference is shouting (all caps) at another group member. In the response, a group regular politely points to the fact that Matt is a newbie in the group and asks him to stop shouting:

Area: FIDO - ST: The Next Generation
Msg#: 673
From: Bill Nichols
To: Matt Oshields
Subj: Cool it, please.

MO>NOW GOD DAMNIT warp 13 is NOT 9.99999... IT'S
MO>13.000000000000000000000000000000000000000000000000

Matt, it's just that you haven't been here very long. You're yelling about a misunderstanding of things that most folks here talked to death quite a good while ago. Those of us who've been here more than a few weeks know what he means, wrong though it might be.
But please do cut out the yelling. Your links are going to get cut if you don't flex some self-control muscles pretty soon. <vbg> This sort of thing is VERY highly frowned upon in Fido.
-!-
þ QMPro 1.53 þ "If you're drunk, I'm funny."

-!- WILDMAIL!/WC v4.11
! Origin: Louisville Hot House (1:2320/180.0)

In this example, a group regulars define the breach of social norms as inadvertent, and work to establish the definition of the guilty party as a newbie--as one who did not know any better. This is typical of how people new to a conference are socialized to the rules. When the rule infraction appears to be unintentional on the part of the newbie, the dialogue and framing of the situation by the regulars begin mildly, with suggestions that the newbie may want to change his or her behavior. If this sort of negotiation does not work, the regulars usually attempt increasingly harsh means of negotiation.

More often than not, newbies openly acknowledge their status as such, thus taking upon themselves the job of defining the resulting communication or situation as one which could be attributable to newbieness, just in case something goes wrong. Depending upon the situation in which the newbie finds him/her self, the level of self-disclosure and admission of newbieness escalates as warranted in an attempt to correct a potentially bad situation. In the posting below, a newbie to the National Cooking Echo conference has previously posted recipes to the group which called for alcohol. Additionally, he also wrote a message which could have been interpreted as an attempt at marketing a product in the conference. He later learned that both of these acts were violations of conference norms, and is attempting to make up for it after the fact by claiming newbieness and acknowledging the ability of the moderator to cut his feed as punishment for such violations:

Area: National Cooking Echo

Msg#: 72
From: Ron Curtis
To: Pat Stockett
Subj: Re:Welcome!

PS> Welcome to the Cooking Echo! It's very nice having you here with us.

Thank you so much for the warm welcome. This is the only echo I know of where the Moderator takes such pains to be friendly.

PS> I'd love to see more of your recipes. The old ones fascinate me. For
PS>brewing spirits you'll need to get a hold of an echo called Zmurgy. The
PS>handle anything that is distilled or brewed. -Pat

Oooops! Wish I had read this yesterday! I posted a couple of wine recipes and they are already winging their way across the Atlantic towards your computer. Could we pretend that they are alcoholic dips this time if I promise not to post any more 'Falling Down Water' recipes %-) I would hate to get a nice welcome and get 'Moderated' all in my first week.

It is possible I may have, unwittingly, fallen foul of one or two rules. I put my mail address in a posting which I am now told is a definite No! No! I may have also given the impression of selling something -Egg Cups- My idea was if someone wanted some I would purchase them and post them on, as long as I was reimbursed the purchase price and postage.

Grovel! Grovel! (Licking mistresses hand) Any violation of the rules was through ignorance of them. I am going to write on the blackboard (I seem to recall you Colonials say "Chalkboard" though I could be, and often am, wrong) 300 times "Put Brain in gear before engaging keyboard".

Being such a warm hearted 'New Joisey' lady I know, or hope, that you will forgive me ;-)

Love from The Limey over the pond!
Ron in cold, cold Blackpool on the NW coast of Ye Merrie Olde England
ron.curtis@bbs.warp.co.uk

... If at first you don't succeed, join the club.
-!- FMail/386 1.02
 ! Origin: Crock's Corner BBS Tel 01253 291023 (2:250/607)

The moderators of Echo conference areas, unlike those in Usenet newsgroups, play an important role in the day-to-day workings of conference life. Like Undernet ops, they are responsible for welcoming newbies and making them feel at home, as well as for apprizing them of rule violations in a polite way. While the group regulars clearly assist in these tasks, the ultimate responsibility for social order lies with the moderator. Individual members of conference areas, while they may choose to skip over messages from people they do not like, have no technical tool like the 'killfile' in Usenet or the "/ignore" command in Undernet with which to avoid annoying behavior. Thus, the actively involved moderator has been developed as a strategy through which regulars may maintain order in their communities.

Of course, it is not only the newbies in a conference who must be reminded of the rules. Besides socialization of newcomers, conferences areas serve as a strategy through which the regulars achieve commit and reciprocation from the other. Any threat to the achievement of this goal is considered annoying behavior, and is grounds for a warning from the regulars or the conference moderator(s). Consider the following example from the UFO Discussion conference area in which one regular challenges the post of another regular. The regulars in this group consider this conference to be a safe haven; a place in which they can discuss events and issues that outsiders would consider 'crazy'. For this reason, the assistant moderator (Glenda Stocks) warns Joe Morris that his attack on a previous post was uncalled for. Even though another regular sides with Joe, the chief moderator of the group (Ron Pappas) has the final word about what is acceptable interaction, and Joe eventually apologizes:

Area: Fido: Open UFO Discussion & Science Thought

Msg#: 27
 From: Glenda Stocks
 To: Joe Morris
 Subj: Latest On Mass Landings From Ashtar.

JM> Gimme a break.... What's next? dancing bears?

Joe,

Please don't make fun of people.

Thanks,
Moderator

-!- GEcho 1.02+

! Origin: FREQ SEARCHNT.ZIP 508-586-6977 SearchNet HQ- (1:330/201)

Area: Fido: Open UFO Discussion & Science Thought Msg#: 103

From: Ron Pappas
To: ALL
Subj: Latest On Mass Landings From Ashtar.

Hi jp,

> > JM> Gimme a break.... What's next? dancing bears?

> >Please don't make fun of people.
> >
> >Thanks,
> >Moderator
> >
> I don't think he's making fun of anyone. I think he's got a pretty valid point.
>There's plenty of proof as to the existence of UFO's, but some of the stuff
>that gets posted to this list is fantasy based on no facts whatsoever. There
>are people who read this list who'll believe anything they are told. I could
>probably make up a pretty decent story that sounds plausible and someone
>would buy into it.

I don't believe it is necessary to belittle or to "poke fun at people" for voicing their opinions. If you disagree on something - then voice your own opinion on the matter with dignity for yourself as well as others.

Peace,

Pap...

The College Board 864.878.7340 FIDO - 1:3639/60

-> Send "subscribe i_ufo-l " to majordomo@world.std.com
-> Posted by: "Ron Pappas" <rpappas@ix.netcom.com>

---SnetMgr 0.60 [r0001]

! Origin: SearchNet HQ BBS (508)586-9404 (1:330/201)

Area: Fido: Open UFO Discussion & Science Thought

Msg#: 149
From: Joe Morris
To: Glenda Stocks
Subj: Latest On Mass Landings From Ashtar.

GS> Please don't make fun of people.

Sorry... It just gets a little thick...

-!- GEcho 1.00
! Origin: The Mosquito Byte - Lacey WA - (360) 459-1892 (1:352/35)

As was the case with Usenet newsgroups, Fidonet conference areas fall along a continuum; some are purely chat oriented, and some are strictly topic-oriented. Regardless, all are places in which individual members achieve a sense of belonging and commitment from others. Depending upon the unique style of the conference, the means by which this goal is accomplished will vary. In those groups which are more chat-oriented, a sense of group belonging and concern for others is achieved through threads in which individuals disclose day to day problems and triumphs in a sequence of short messages. In order to conform to the 'chat' nature of these conferences, message style emphasizes brevity; just enough detail is included to get the point across. In the posting below, a regular member of the National Cooking Echo conference who had previously thought her BBS (and thus her access to the conference) might be going out of business informs other regulars that the BBS will remain up. The regulars respond with congratulations, pointing out that they consider her a friend, and would miss her, were her access to the group discontinued:

Area: National Cooking Echo

Msg#: 171
From: Peg Dietrich
To: Iris Grayson
Subj: Announcement

Socialization and Personal Identity

-=> Quoting Iris Grayson to Joel Ehrlich <=-

IG> i'm writing this announcement to you prematurely--just prior to receiving
>my IG sysop's confirmation--he told me he'd official announce in today's
>download IG that the bulletin board will remain up.

Yeah! I consider you my friend, and I am glad that you aren't leaving.

 Peg
... Because of us, America works!
___ Blue Wave/QWK v2.12
-!- FMail/386 1.02
! Origin: Vickie's Palace - Leslie, MI - (517)589-5954 28.8k (1:2330/200)

Area: National Cooking Echo

Msg#: 185
From: Iris Grayson
To: Pat Stockett
Subj: Announcement

PS> I'm so glad the bbs is staying up. Tell him or her thank you for us. :)

oh, believe me--i already have (and backed it up with a check).

PS> I hope your wrists heal soon. And don't give the capitals another thought.
:)

thanks, pat...i appreciate your good wishes.

 * SLMR 2.1a * Shin: a device for finding furniture in the dark.
-!- GEcho 1.11+
! Origin: (1:352/256)

Area: National Cooking Echo

Msg#: 194
From: Iris Grayson
To: Joel Ehrlich

Subj: Announcement

JE> IG> i don't have to find a new home. hooray. (i do have to type without
JE> HOORAY!!!!
JE> Score another big one for the home team!
JE> Great news, Iris.

[ala Billy Crystal as Fernando] you sure know how to make a gal feel great, dahling! :-)

* SLMR 2.1a * Saw a car so decrepit, called it "Puff, the Tragic Wagon"
-!- GEcho 1.11+
! Origin: (1:352/256)

 In chat-oriented groups like this one, regulars work to integrate a display of personal concern for others into the general format of the group, which emphasizes light conversation in a format conducive to the conservation of bandwidth.
 Contrast a chat-oriented conference with those that are designed to discuss a specific topic. While one might think that a group designed to cover a particular topic would frown upon extensive threads that could be classified as 'irrelevant', the opposite turns out to be true, at least with regard to messages concerning personal disclosure and group cohesion. The UFO Discussion conference is a group which falls in the middle of the topic-chat continuum, and combines the format of chat-oriented groups with that of topic-oriented groups. It is the format, in part, which determines what types of messages are defined as acceptable topics of conversation by members:

Area: Fido: Open UFO Discussion & Science Thought

Msg#: 82
From: M.A.F. Nienaber
To: ALL
Subj: California Quakes

Dear Californians,
It must be horrible to live with those predictions. I pray that nothing bad will happen to you. Life would be sad and drab without you all. Merry Christmas

and Happy New Year. May you all remain in good health, with roofs over your head and power and water intact. Love,
Maria

Name: M.A.F. Nienaber
E-mail: nieuwenh@pop.pi.net (M.A.F. Nienaber)
Date: 08/12/95
Time: 09:23:47

This message was sent by Chameleon

-> Send "subscribe i_ufo-l " to majordomo@world.std.com
-> Posted by: "M.A.F. Nienaber" <nieuwenh@pi.net>

---SnetMgr 0.60 [r0001]
 ! Origin: SearchNet HQ BBS (508)586-9404 (1:330/201)

Area: Fido: Open UFO Discussion & Science Thought

Msg#: 83
From: John F. Winston
To: M.A.F. Nienaber
Subj: California Quakes

Now to answer a person.
John Winston.
On Mon, 25 Dec 1995, M.A.F. Nienaber wrote:
> Dear Californians,
> It must be horrible to live with those predictions.
JW Thanks for you concern. I now live above 2,000 feet elevation, (2,750 feet to be exact) and some people say I moved up here to get away from the San Francisco Bay Area to get away from the Earthquakes. I would like to state that that is not true but it is a good idea. We even had a 4.6 earthquake up here the other night and Yogi and I didn't even feel it.
John Winston.

-> Send "subscribe i_ufo-l " to majordomo@world.std.com
-> Posted by: "John F. Winston" <johnfwin@mlode.com>

---SnetMgr 0.60 [r0001]
! Origin: SearchNet HQ BBS (508)586-9404 (1:330/201)

Area: Fido: Open UFO Discussion & Science Thought

Msg#: 87
From: Steve Wingate
To: you
Subj: California Quakes

> Dear Californians,
> It must be horrible to live with those predictions.

Not really. I have been hearing for years that California is going to fall off into the ocean. Oh well, the weather has been really nice so far. But I have my raft already inflated just in case <g>.

> I pray that nothing bad will happen to you. Life would be sad and drab
>without you all.

Thank you for your prayers.

> Merry Christmas and Happy New Year. May you all remain in good health,
> with roofs over your head and power and water intact.

After the recent storms which wiped out the power to over 800,000 people here, I really appreciate these things.

Steve

Love,
> Maria

```
         ^
     <<<<<<|>>>>>>
    <<<<< steve@linex.com >>>>>
   <<<<<<< http://www.linex.com/ufo >>>>>>>
  <<<<<<<<< Anomalous Images and UFO Files >>>>>>>>>
```

```
  <<<<<< Citizens Intelligence Access BBS 415.927.2435 >>>>>
  <<<<<<<<<<<<<<<<<<<<<<<<<<<<<<<<*>>>>>>>>>>>>>>>>>>>>>>>>
  -> Send "subscribe i_ufo-l " to majordomo@world.std.com
  -> Posted by: Steve Wingate <steve@linex.com>

  ---SnetMgr 0.60 [r0001]
   ! Origin: SearchNet HQ BBS (508)586-9404 (1:330/201)
```

Area: Fido: Open UFO Discussion & Science Thought

Msg#: 91
From: Kwinters
To: M.A.F. Nienaber
Subj: California Quakes

> Dear Californians, It must be horrible to live with those predictions.
> I pray that nothing bad will happen to you. Life would be sad and drab
> without you all. Merry Christmas and Happy New Year. May you all
> remain in good health, with roofs over your head and power and water
> intact.
>Love,
> Maria

Thank you for your kind wishes and prayers, Maria. They are much appreciated. Actually, although we know that a large quake could happen anytime, just by virtue of where we live, most people do not spend much time dwelling on it. In fact, I've never heard another person discuss the Scallion predictions outside of the Internet. I don't take them too seriously, myself, but am prepared for big quakes nonetheless. And the geologists I've spoken to say there's zero chance that California will break off and fall into the sea - it's just physically impossible given the underlying geological structures.
But we will likely receive an 8 pointer sometime in the next 30 years - if history is any indicator - so we all try to ready.

Thanks again,

Karen

-> Send "subscribe i_ufo-l " to majordomo@world.std.com
-> Posted by: gwprod@primenet.com (Kwinters)

---SnetMgr 0.60 [r0001]
! Origin: SearchNet HQ BBS (508)586-9404 (1:330/201)

In this series of four posts, the format through which the regulars display concern for the other resembles that of the chat-oriented groups. Messages are relatively brief, and the thread is not one which continues in the conference for any length of time. Compare this with the following posts from a group that lies primarily toward the topic-oriented end of the continuum.

The example below (in an abbreviated form) is one from the NFL conference area. The regulars of this group are engaged in an endless stream of discussion and debate regarding favorite teams, recent games, and NFL policy in general. Like the talk.origins newsgroup in Usenet, this NFL conference is one which limits (and basically excludes) any discussions of a personal nature. Rather than establishing a sense of self and belonging through sharing of personal items, members establish group cohesion through the debate-style cycle of statement and response. However, in keeping with the generally relaxed atmosphere of Fidonet, these posts are sprinkled with polite words and emoticons, which allow the cultural framework of friendliness to be continually reestablished:

Area: FIDO - National Football League

Msg#: 2
From: Doug MacFarlane
To: John Mitchell
Subj: Cowboys defense

JM> -- This was in a message from Doug MacFarlane to Dan Dyer --

DM> I've seen him make some bad calls, but, Cowboy fans will never be
DM>satisfied with him until he wins the SB. Sometimes I think the fans
DM>expect too much from him.....who gets the credit for the 35-0 blowout
DM>against the Giants?

JM> What 35-0 blowout? On December 17, I saw a narrowly-eked-out

JM> one-point win.

It was the first MNF game of this season......Emmitt's *FIRST* run was a TD, something like 60 yards --- as a 9'er fan watching his team struggling to beat the Saints on opening day of the season, well, that Cowboy game had me scared!

Since then, I've finally learned the lesson to not count any team in or out until about the 2nd or 3rd week in December....

Thanks,

Doug MacFarlane

-!- Maximus/2 3.00
! Origin: The Rock BBS--The Vikings *mean* Victory! (1:387/31)

Area: FIDO - National Football League
Msg#: 51
From: John Mitchell
To: Doug MacFarlane
Subj: Cowboys defense

-- This was in a message from Doug MacFarlane to Dan Dyer --

DM> I've seen him make some bad calls, but, Cowboy fans will never be
DM> satisfied with him until he wins the SB. Sometimes I think the fans
DM >expect too much from him.....who gets the credit for the 35-0 blowout
DM>against the Giants?

What 35-0 blowout? On December 17, I saw a narrowly-eked-out one-point win.

-!-
! Origin: We're having Post FUN!! in KC at 816-942-5641 (1:280/103)

Area: FIDO - National Football League

Msg#: 52
From: Terry May
To: Doug MacFarlane
Subj: Cowboys defense

Re: _Cowboys defense_, Doug MacFarlane wrote to Dan Dyer on 16 Dec 95:

DM> I'll be the weird one on the echo and say that I don't think it was that
DM>bad of a call --- the only thing that bothered me was that they were tied,
DM>and in bad field position to do it.

THAT'S THE POINT! No one is suggesting that you should never go for it on 4th down. What people are saying is you don't go for it on 4th down on your own 29 in a tie game with two minutes left. Oh, and don't forget that HFA was on the line.

That's like calling a QB sneak on 3rd down and you say you have no problem with the call, except there were 10 yards to go for a 1st down.

DM> But, if Emmitt would have broken thru the line, and made like a 30 yard
DM> run, we all would of heard how great the Cowboys were in that game,
DM> with little or no credit going to Switzer....but, since it didn't go that way,
DM> Switzer becomes the scapegoat.

No, if it would have worked, we'd simply be saying Switzer got away with making a bonehead call. It was a bonehead call whether or not it worked. It wasn't a bonehead call just because it failed. It was a bonehead call because the risk far outweighed the reward.

... NFL Week 14 Rushing Totals: Marcus Allen 124 - Oakland Raiders 8!
-!- JetMail 1.00alpha
! Origin: *[Rebel BBS]-[Las Vegas]-[HST/V32b]-[702/435-0786]* (1:209/745)

Area: FIDO - National Football League

Msg#: 128
From: Will Morgan
To: Doug MacFarlane
Subj: Cowboys defense

DD> Going on 4th down from your own 29 yard line with the game tied and 2
DD> minutes to go is tantamount to suicide.

DM> I'll be the weird one on the echo and say that I don't think it was that
DM> bad of a call --- the only thing that bothered me was that they were tied,
DM> and in bad field position to do it. But, if Emmitt would have
DM> broken thru the line, and made like a 30 yard run, we all would of heard
DM> how great the Cowboys were in that game, with little or no credit going
DM> to Switzer....but, DM> since it didn't go that way, Switzer becomes the
DM> scapegoat.

During the Giants game Sunday, we had two fourth & short situations deep in Giants territory. Switzer went for field goals in both cases. He was soundly booed both times.

Bottom line, we won the game (a MUST win), although just barely.

... Dallas Cowboy Fanatic's Companion: Coming Soon for Windows/Win95!

-!- PPoint 1.96
! Origin: DALLAS COWBOYS: 4 Time World Champions! (1:3819/128.103)

Area: FIDO - National Football League

Msg#: 151
From: Skywalker
To: Doug Macfarlane
Subj: Cowboys defense

DM> I've seen him make some bad calls, but, Cowboy fans will never be
DM> satisfied with him until he wins the SB. Sometimes I think the fans

DM>expect too much from him.....who gets the credit for the 35-0 blowout
DM>against the Giants?

I do not envy Switzer at all. He had the enviable task of taking over for a popular two time SB winning coach, who was fired unexpectedly. He was thus hired for the worse job in the NFL. If he wins, it's because of the job JJ did the year before. If he looses, it's his fault. He even had it worse than Siefert, at least his arrival was somewhat expected/planned. Switzer arrived on a bad scene, and regardless of what he did (unless he wins the big one) is be considered a failure... while JJ will always be remembered as having won two superbowls then being fired.
Porter Broyles

-!- GEcho 1.11+
! Origin: Trantor BBS 303-670-7947 670-5871 (1:104/356)

The final two strategies for achieving social order and legitimation of the self in the Fidonet community are the BBS log-in registration requirements and Echo conference message formats. Fidonet Echo conference messages are very similar in format to those found in Usenet newsgroups; all messages begin with header information regarding the identities of the author and recipient(s) of the message, message subject, and so forth. Further, conferences messages generally conclude with additional identity-oriented information, such as the BBS on which the post originated (which enables it to be tracked, if it is found to be in serious violation of a Fidonet rule) and the sender's Fidonet netmail address. And, as was the case with Usenet messages, the texts of Fidonet messages are formatted such that quoted material is clearly marked with both the individual's initials and the '>' sign. Each of these strategies lend order to the interaction within conference areas by providing information which legitimates who said what to whom and in what context.

Unlike Usenet, however, the Fidonet community has an additional source of identity information about members; Fidonet BBSs require each new member to register a legitimate identity with the operator of the board. Each time an individual logs into a new BBS for the first time, he or she is required to give the sysop certain personal information. In general, the sysop requires technical information concerning the type of computer the new member is using (what languages it is capable of transmitting, what type of terminal it is, and whether or not it can support graphics files) and personal information about the user (full name, password, address, phone number, and in some cases even sex and birthday).

Such information enables the sysop and conference moderators to work together in order to maintain social order within the Fidonet community in two ways. First, the technical information allows sysops to stay current on the needs and abilities of users in order to maintain the efficiency of the system. Second, the personal information enables the sysop and moderator to control potentially deviant behavior both before and after the fact. All individuals understand upon logging in to the system that other community members are capable of utilizing log in information to define their identities. This ability may serve to limit deviant and potentially threatening behaviors at the outset, and certainly allows for community members to define deviant behavior after the fact and punish the guilty party accordingly.

Echo conference Rules, the conferences themselves, moderators, message formats and BBS registration requirements all play important roles in the maintenance of social order and personal identity throughout Fidonet. Fidonetters pride themselves on a relaxed and friendly system of communication through which people can meet with others to discuss similar interests. The institutions and groups which they have developed serve those goals by socializing newbies and ensuring the integrity of individual and newsgroup identities. Through such institutions and groups, Fidonetters maintain the beliefs and system of values which are the foundation of Fidonet.

Though there is always the potential for disruption, there is a very high degree of order and stability within and across conference areas. Conference names provide a general frame within which interaction is defined, and the group moderator(s) and regulars, through a complex system of interaction, work to maintain order in their groups. When the situation warrants, Fidonetters engage in and respond to a very high degree of self-disclosure; this sort of investment and legitimation of the self serve to promote the feeling of membership. Fidonetters are just like Usenetters and Undernetters; they are committed to their community identities and work to protect them from those who threaten the social order.

ENDNOTES

[1] This distinction between intentional and unintentional newbie behavior is discussed in Chapter Six.

[2] The reader will note also that the format of the private channel differs in other respects from the public channel. Most notably, the dialogue is more in-depth; longer lines of text are accepted and several lines of text in a row by one individual are allowed in order that the person may get the complete idea across.

[3] One should also note that within Cici's response, she brings up two other relevant Usenet norms: 1) if one is going to cite statistics or other facts, one must be prepared to cite the specific source of those facts for others so that they may verify them; and 2) one should not waste time and bandwidth in allowing a thread to escalate beyond its usefulness.

[4] There are instructions available for Usenetters who wish to post anonymously, but such posts are only generally accepted when other group members feel that the content of the post justifies the anonymity. For example, someone posting to alt.support.depression may be discussing issues of a personal nature that he or she would not want others around him/her to know about.

CHAPTER 6

DEVIANT BEHAVIOR: ITS DEFINITION AND CONTROL

Often times, one's clearest understanding of social order stems from the observance of the violation of that order, social deviance. That is, we are most aware of social rules and values when we witness their violation. Discussions of this social order usually suggest that the opposite of order is chaos; that the absence of social order means that actors act without taking others into account, thus making cooperation impossible. They further suggest that, where social order exists, actors know what to do; they act with each other in mind and according to the social patterns developed over time. Within this framework, deviance is seen as a breakdown (if only temporarily) of the social order; something has gone wrong with the organizational structure of society.

In general, deviance refers to violations of social norms. Norms are behavioral codes, or prescriptions, that guide people into actions and self-presentations that conform to social acceptability. These behavioral codes may be placed into three categories: folkways, mores and laws. Folkways are simple everyday norms based on custom, tradition or etiquette, and their violation does not generate serious outrage. Mores are based on broad societal morals, whose infraction generates more serious social condemnation; upholding mores is seen as 'critical' to the fabric of society. Laws are the strongest norms because they are codified by social sanctions, and the people who violate them are subject to arrest and punishment.

However, the decentering of physical place (and physical presence), combined with the fact that all social norms are written down in a text-based reality, calls into question the idea of categorizing these norms by degree of formality. In computer-mediated communities, there are not necessarily 'laws' the way we generally understand them. This does not mean, though, that members have no means of social control. What it does mean is that the importance of folkways and mores in social order needs to be re-examined. Social order and social control are established and maintained in the course of everyday interaction. If folkways and mores are understood as the common, everyday norms that make up the 'fabric of society', then their enforcement through such everyday interaction is what is critical to the maintenance of order, not reference to 'laws' as such:

> "All social groups make rules and attempt, at some times and under some circumstances, to enforce them. Social rules define situations and the kinds of behavior appropriate to them, specifying some actions as 'right' and forbidding others as 'wrong'. When a rule is enforced, the person who is supposed to have broken it may be seen as a special kind of person, one who cannot be trusted to live by the rules agreed on by the group. He is regarded as an outsider." [Becker 1963, 1]

These social rules may be formally enacted into law, or they may be informal agreements, but such a distinction is relatively unimportant. What is important, in the study of deviance, is that one pay attention to the 'operating rules' of groups; those kept alive through attempts at enforcement (by any means). Further, whether an act is deviant depends on how other people react to it; the response of the 'other' is what is problematic. Variations in response to particular actions occur over time, and depend as well on who commits the act and who feels he has been harmed by it; rules are applied more to some persons than to others. Thus, deviance is not a quality that lies in behavior itself, but in the interaction between the person who commits an act and those who respond to it. Because human action is normative, human values determine what is to be considered a 'problem'.

Howard Becker's approach puts deviance outside the individual, rather than inside, and claims that categorization is not an automatic process, but one involving social power and negotiation of the definition of the situation. While he does not explicitly discuss what grounds people use to attempt to apply the label 'deviant', others have suggested, through the analysis of his typology of deviance[1], that deviant behavior is not necessarily that which violates social norms. Rather, deviant behavior is that which is perceived as a threat to a particular group's values, within a particular situation. [Lauer and Handel 1983; Hewitt 1984] Deviance, a perceived threat to social values, is also a threat to the identities of individuals involved; one constructs who one is through the values to which one subscribes.

What this means in terms of the place of deviance in society, then, is that deviance does not mark the breakdown of social order. Instead, it marks the specific points at which one may most fully understand the social order. Deviance is constructed as a means of pointing out what the social rules are and the consequences for violating those rules. Constructing deviance is the 'politics of reality' [Hewitt 1984, 238]:

> "The 'reality' of deviance can only be achieved when people are imagined and successfully labeled as such. Deviance. . .refers to the socially constructed negative moral meanings that are situationally generated to describe behavior and personal attributes perceived as different and disturbing to certain audiences. The unifying factor among all behaviors and attributes named and categorized as deviant is that they are perceived to be at variance with some group's definition of what is preferable or morally acceptable. To create 'deviance' is simultaneously to create 'normality'; we can know one only in relation to the other." [Pfuhl and Henry 1993, p. 22]

This process is no different in CMC communities than it is in face-to-face communities. The violation of the everyday life rules of netiquette are precisely those which generate 'serious outrage' when violated.

THE INTERNET RELAY CHAT COMMUNITY

IRC supports mechanisms for the enforcement of acceptable behavior. Channel operators (ops) have access to the /kick command and the /ban command. . . .participants can be 'kicked'. . .for being obnoxious. Obnoxiousness seems a somewhat trivial term to warrant the use of such textually violent commands such as /kick and /ban. The word trivializes the degree to which abusive behavior, deceit and shame can play a part in interaction in IRC. . . Violators of the integrity of the IRC system are marginalized, outcast, described so as to seem insignificant, but their potential for disrupting the IRC community is suggested by the emotive strength of the words with which they are punished... [Reid 1991, 18]

Chapter Five pointed out that newbies may unknowingly violate any number of social norms, and suggested that this type of behavior is only defined as problematic, or deviant, by the regulars if it is seen as deliberate. This definition depends largely upon the extent to which the newbie is willing to: 1) acknowledge newbie-ness, 2) apologize, and 3) accept assistance or guidance from others. All regulars on IRC were once newbies themselves; they anticipate and understand that everyone who joins the Undernet must go through a learning process, and that it is part of their job to help them. However, when it appears that the individual is unwilling to accept help or alter his or her behavior, and when he or she too bluntly challenges the regulars on their definitions of the situation, this is cause for being labeled as deviant. Any number of repercussions may result.

Newbies are warned, in almost every channel, that they are currently in the friendliest channel, and that it is the other channels in which deviants are to be found. Each group considers itself to be a very cohesive unit in which the members 'stick up for' each other and work to keep their channel orderly and friendly. Often times a core group of the regulars will serve as what Becker [1963] referred to as 'moral entrepreneurs'; groups of observers, watchdogs and whistle blowers which 'lay people' rely on in complex situations to define and deal with deviant behavior:

[DISP-#geezers][miranda]   oh! that is one thing of interest...the core people REALLY do stick together in a way that is pretty unusual...
[DISP-#geezers]<Marcia_Br>   how do you see them sticking together?
[DISP-#geezers][miranda]   say one of the women is being harassed by a new male...the others, men and women both, deal with it...usually immediately
[DISP-#geezers]<Marcia_Br>   oh. that's good. helping each other i mean.
[DISP-#geezers]<Marcia_Br>   so there really is a sense of belonging to a group and sticking up for each other.
[DISP-#geezers][Heatgain]   like an ant colony....

[DISP-#geezers][miranda] or in the men's case...if some woman is being ...idiotic? the women are usually pretty quick to make it rather clear...oh it's GREAT
[DISP-#geezers][miranda] ROFL
[DISP-#geezers]<Marcia_Br> hehehe
[DISP-#geezers]<Marcia_Br> what do the women do to stick up for the men?
[DISP-#geezers][miranda] smiling...that is a good question and hard to explain...thinking about serena and ie heat...
[DISP-#geezers][Heatgain] i don't think it's a matter of the women sticking up for the men as much as it's a matter of the women banding together AGAINST an intruder..
[DISP-#geezers][Heatgain] i might be wrong...
[DISP-#geezers][miranda] yes...
[DISP-#geezers]<Marcia_Br> ohhh. now that's even more interesting!
[DISP-#geezers][Heatgain] <---stepping carefully..
[DISP-#geezers][miranda] no...I think that is true...
[DISP-#geezers]<Marcia_Br> how do you know who an intruder is?
[DISP-#geezers]<Marcia_Br> they are just rude or something?
[DISP-#geezers][miranda] the women on the channel who are close...REALLY care about each other...but if they don't like a woman...she's dead
[DISP-#geezers][miranda] heat...are we rude???
[DISP-#geezers]<Marcia_Br> no i mean is the intruder rude. hehehe
[DISP-#geezers][Heatgain] yes...the women can be cruel to a newbie woman..
[DISP-#geezers]<Marcia_Br> what do they do??
[DISP-#geezers][Heatgain] well...they would make comments on what she just typed....they will bandy about initials, meanings known only to certain people..
[DISP-#geezers][miranda] marcia it's the....no...usually it's a woman who comes in looking for net sex...or is just...very foolish..tho, we can certainly be fooish
[DISP-#geezers][miranda] foolish
[DISP-#geezers][miranda] marcia...in a lot of ways...41plus is matriarchal
[DISP-#geezers][miranda] :)

[DISP-#geezers][miranda] it defies almost every stereotype concerning computers and their use
[DISP-#geezers]<Marcia_Br> yeah.
[DISP-#geezers][miranda] on the other hand...when a woman is well received...she is like a new sister...
[DISP-#geezers]<Marcia_Br> so, how does one go about being well-received?
[DISP-#geezers][Heatgain] yes....sister is a word the women use to each other....the core of regulars..
[DISP-#geezers][miranda] I think the well-received women on the channel are usually very bright...very articulate...and NOT looking (instantly or at least crudely) for a man
[DISP-#geezers][Heatgain] that could go for the men too..
[DISP-#geezers]<Marcia_Br> so, when a newbie shows up, how long does she have to prove she's bright and articulate?
[DISP-#geezers][miranda] about five seconds
[DISP-#geezers]<Marcia_Br> lol
[DISP-#geezers][miranda] :)
[DISP-#geezers]<Marcia_Br> lol
[DISP-#geezers]<Marcia_Br> still
[DISP-#geezers]<Marcia_Br> yeah, otherwise, the women in the group will think she's trying to pull something.
[DISP-#geezers]<Marcia_Br> if she only pays attention to the guys i mean.
[DISP-#geezers][miranda] if that is all she does...she usually is pretty obvious right away...

This is an accurate description of how most of the channels in the Undernet work; a core of regulars works to maintain the atmosphere of the channel. They seek, through a process of negotiation with other regulars and the individual in question, to define particular types of behavior as violations of Undernet norms and punish that behavior in an acceptable manner. Recall the basic IRC rules of netiquette: 1) do not 'flood' the channel with text; 2) do not use beeps in messages; 3) do not use profanity in public messages; and 4) do not harass other users with unwanted messages or comments.

Depending on how the channel regulars define the severity of the violation of these norms, the 'appropriate' punishment will vary. Chapter Five dealt mainly with newbie behavior which was defined as 'normal' within the course of socialization. Here we are dealing with behavior which has been defined as deviant because it is seen, given the particular circumstances, as intentional-- as behavior that no 'true' newbie would think of (or be capable of) engaging in. In such instances, means of social control move beyond those discussed in Chapter Five into the realm of /kicks and /bans. However, even here there is negotiation involved. The amount of time an individual is given to correct him-or herself before being kicked, whether or not to allow a kicked (or banned) individual back into the channel, and the harshness of the kick (ranging from the simple /kick to the elaborate 'splatter kick') are all points that are negotiated in interaction.

Regarding the first two rules against flooding and beeping, consider the following examples. In the first case, Sub-Zero repeatedly enters the channel #41plus and directs a mini-flood toward the women in the channel. Because Sub-Zero only sent a mini-flood, one which did not disrupt the connections of the channel members, he was eventually given a simple /kick from the channel:

[DISP-#41plus][SERVER] Sub-Zero!~alexb@sluggo.pcdocs.com has joined this channel
[DISP-#41plus][Sub-Zero] Smoooches with
[DISP-#41plus][Sub-Zero] Pre-heated electric lips to
[DISP-#41plus][Sub-Zero] all the cyberbabes! From
[DISP-#41plus][Sub-Zero] Antarctica!
[DISP-#41plus][Sub-Zero] Smoooches with
[DISP-#41plus][Sub-Zero] Pre-heated electric lips to
[DISP-#41plus][Sub-Zero] all the cyberbabes! From
[DISP-#41plus][Sub-Zero] Antarctica!
[DISP-#41plus][SERVER] Sub-Zero has left this channel
[DISP-#41plus][ACTION] Gyrfalcon readies a Pre-heated kick....
[DISP-#41plus][ie] iris you cyberbabe you
[DISP-#41plus][SERVER] Sub-Zero!~alexb@sluggo.pcdocs.com has joined this channel
[DISP-#41plus][Sub-Zero] Smoooches with
[DISP-#41plus][Sub-Zero] Pre-heated electric lips to
[DISP-#41plus][Sub-Zero] all the cyberbabes! From
[DISP-#41plus][Sub-Zero] Antarctica!
[DISP-#41plus][Sub-Zero] Smoooches with

[DISP-#41plus][Sub-Zero]   Pre-heated electric lips to
[DISP-#41plus][Sub-Zero]   all the cyberbabes! From
[DISP-#41plus][Sub-Zero]   Antarctica!
[DISP-#41plus][Sub-Zero]   Smoooches with
[DISP-#41plus][Sub-Zero]   Pre-heated electric lips to
[DISP-#41plus][Gyrfalcon-KICK]   has kicked Sub-Zero from #41plus Gyrfalcon

While this type of flooding activity is an annoyance, it does not jeopardize the stability of the channel connections, nor does it severely impair the ability of others in the channel to communicate. The perpetrator is dealt with through sarcastic references to 'cyberbabes' and is given a simple /kick. Contrast this with the example below, in which TekLord enters the channel and begins a much more severe flood:

[DISP-#gaychat][SERVER]   TekLord!~ircuser@tesla.pirate.org has joined this channel
[DISP-#gaychat][PoPaToP] [TekLord] (via a Public IRC Client)
[DISP-#gaychat][TekLord]   ^ V^ e^ v^ e^ S^ • V• e• v• e• S• • ^ V^ e• ^ v^ e• ^ S^ •   VeveS ^ V^ e• ^ v^ • e^ S^ ^ V^ e^ v^ e^ S^ • V• e• v• e• S• • ^ V^ e• ^ v^ e• ^ S^ •   VeveS ^ V^ e• ^ v^ • e^ S^ ^ V^ e^ v^ e^ S^ • V• e• v• e• S• • ^ V^ e•

[DISP-#gaychat][TekLord]   ^ ^• [• ^• [• ^• [• ^• [• ^• [• ^• [• ^•
[• ^• [• ^• [• ^• [• ^• [• ^• [• ¨ ^• [• ^• [• ^• [• ^• [• ^• [• ^• [•
^• [• ^• [• ^• [• ^• [• ¨ ^• [• ^• [• ^• [• ^• [• ^• [• ^• [• ^• [• ^•
[• ^• [• ^• [• ^• [• ^• [• ^• [• ^• [• ^• [• ¨ ^• [• ^• [• ^• [• ^• [• ^• [•
[DISP-#gaychat][TekLord]   ^ ^• [• ^• [• ^• [• ^• [• ^• [• ^• [•
^• [• ^• [• ^• [• ^• [• ^• [• ^• [• ^• [• ^• [• ^• [• ^• [• ^• [•
^• [• ^• [• ^• [• ^• [• ¨ ^• [• ^• [• ^• [• ^• [• ^• [• ^• [•
[DISP-#gaychat][TekLord-NOTICE]
   ^ ^• [• ^• [• ^• [• ^• [• ^• [• ^• [• ^• [• ^• [• ^• [•
^• [• ¨ ^• [• ^• [• ^• [• ^• [• ^• [• ^• [• ^• [• ^• [• ^• [• ^•
[• ¨ ^• [• ^• [• ^• [• ^• [• ^• [• ^• [• ^• [• ^• [• ^• [• ^• [•
[DISP-#gaychat][Kyle]   lovely...
[DISP-#gaychat][TekLord-NOTICE]   ^ V^ e^ v^ e^ S^ • V• e• v
• e• S• • ^ V^ e• ^ v^ e• ^ S^ •   VeveS ^ V^ e• ^
v^ • e^ S^ ^ V^ e^ v^ e^ S^ • V• e• v• e• S• • ^ V^ e• ^ v^

e• ^ S^ • VeveS ^ V^ e• ^ v^ • e^ S^ ^ V^ e^ v^ e
^ S^ • V• e• v• e• S• • ^ V^ e• ^ v^ e• ^ S^ • VeveS
[DISP-#gaychat][ACTION]	Ragtop dumps a can of gasoline on TekLord
[DISP-#gaychat][ACTION]	Ragtop lights a match and tosses it.
[DISP-#gaychat][ACTION]	Ragtop laughs as TekLord turns, squirms and smokes to a blackened crisp.
[DISP-#gaychat][Ragtop-KICK]	has kicked TekLord from #gaychat Ragtop
[DISP-#gaychat][SERVER]	TekLord!~ircuser@tesla.pirate.org has joined this channel
[DISP-#gaychat][PoPaToP] [TekLord] (via a Public IRC Client)
[DISP-#gaychat][Ragtop]	It's April, the season of growing, and the fucking weeds are popping up everywhere
[DISP-#gaychat][ACTION]	Ragtop spots a weed in #gaychat
[DISP-#gaychat][ACTION]	Ragtop calls the Orkin man
[DISP-#gaychat][ACTION]	Ragtop sees the Orkin man spray TekLord
[DISP-#gaychat][Ragtop-KICK]	has kicked TekLord from #gaychat Ragtop
[DISP-#gaychat][Ragtop]	psssttzz.., the weed is gone
[DISP-#gaychat][ACTION]	Ragtop mails TekLord the bill.
[DISP-#gaychat][SERVER]	TekLord!~ircuser@tesla.pirate.org has joined this channel
[DISP-#gaychat][Ragtop-KICK]	has kicked TekLord from #gaychat Ragtop
[DISP-#gaychat][Ragtop-MODE]	Has changed TekLord!*ircuser@tesla.pirate.org's mode to +b
[DISP-#gaychat][Hotboy]	that was fun!
[DISP-#gaychat][Scorpion]	everyone's screen mess up?
[DISP-#gaychat][Scorpion]	with beeps?
[DISP-#gaychat][Ziller]	i had beeps....
[DISP-#gaychat][Steven]	Here too.
[DISP-#gaychat][Ragtop]	ziller, you just missed my splatter kicks in action
[DISP-#gaychat][Ziller]	rags, no, i saw
[DISP-#gaychat][Ragtop]	hehehehe

In this particular circumstance, it is clear that TekLord must be an experienced IRCer, otherwise he could not construct such a flood message with beeps. It is this type of flood that can destroy the computer connections of users and make orderly dialogue impossible to achieve. Additionally, TekLord had his computer set in the automatic re-join mode, meaning that immediately upon being kicked from a channel, he would return; another clear indicator of intent to cause harm to the channel.

Because of this, he is confronted by RagTop, one of the channel ops, with a series of 'splatter' kicks. Splatter kicks are designed as much to amuse the channel regulars and create a sense of group cohesion as they are to point out the severity of the norm violation to the perpetrator. It is customary after a splatter kick for channel members to comment on the quality of the splatter. The splatter kick tends to grab the attention of any channel members still engaged in a parallel conversation and in this way serves to draw attention to the consequences of deviant behavior; any channel members who were not 'paying attention' are compelled to ask what the individual did to deserve the kick.

In general, profanity is another type of behavior that is unacceptable in the public channels of the Undernet, and individuals engaging in it receive an immediate reprimand from others, usually ending in a /kick:

[DISP-#chatzone][sensii]   fuck face
[DISP-#chatzone][Emily2-KICK]   has kicked sensii from #chatzone ^ l Emily l^
[DISP-#chatzone][lulu] Sensii.......???????
[DISP-#chatzone][Nessie]   senssi...Be polite please !
[DISP-#chatzone][sensii]   sorry Emily2.
[DISP-#chatzone][Emily2]   sensii ok

In this example, sensii is immediately kicked for inappropriate behavior. She or he returns to the channel and apologizes, and Emily2 accepts the apology. In the situation below, lensmann violates the no-profanity norm, and RechARGE is congratulated on the speed with which he kicked lensmann. Moments later, lensmann returns and pretends not to understand what he did. However, the regulars will have none of that; they have already defined this specific instance as one in which no 'good' person would even unknowingly engage:

[DISP-#35plus][lensmann]   michael jackson is a nigger

Deviant Behavior: Its Definition and Control

[DISP-#35plus][Ali3]   lensmann take it elsewhere
[DISP-#35plus][zenia]   RechARGE: I am a little ahead of you
[DISP-#35plus][RechARGE-KICK]   has kicked lensmann from #35plus ^ I GargOyle I^
[DISP-#35plus][vacutaine]   oops beat me to it
[DISP-#35plus][RechARGE]   :)
[DISP-#35plus][vacutaine]   you missed the recharge hoof of lensmann
[DISP-#35plus][klaatu]   i expected it soon....
[DISP-#35plus][klaatu]   good stuff, recharge..!
[DISP-#35plus][RechARGE]   tanx :)
[DISP-#35plus][Ali3]   weird one........
[DISP-#35plus][SERVER]   lensmann!sigvald@port4.intercom.no has joined this channel
[DISP-#35plus][IceNight]   Thats why I got sent to Antarctica...for not being politically correct...
[DISP-#35plus][RechARGE]   watch the lingo lensmann
[DISP-#35plus][vacutaine]   no n words lens or you're toast
[DISP-#35plus][lensmann]   ?
[DISP-#35plus][RechARGE]   don't act innocent
[DISP-#35plus][lensmann]   ok
[DISP-#35plus][lensmann]   hello
[DISP-#35plus][vacutaine]   hello
[DISP-#35plus][RechARGE]   Lensmann, you cant talk like that on this channel, ok
[DISP-#35plus][lensmann]   ok
[DISP-#35plus][RechARGE]   cool

The fourth basic rule of IRC interaction is that one should not commit 'general harassment', meaning one should not send unwanted messages and/or comments to others, through either public or private channels. The phrase 'unwanted messages and/or comments', however, can encompass a wide array of behaviors, some more or less threatening to social order than others. Consequently, there exist a myriad of ways in which individuals deal with this problem, depending again on how the situation is defined.

For example, there are a number of channels which have either a specialized bot or a modified X bot, which is intended as a means of social control. For example, if one joins the channel #bisex, one is greeted by this auto-op message from BibotII:

[DISP-BibotII][BibotII]   Hi Marcia_Br! Welcome to Bisex
[DISP-BibotII][BibotII]   Have fun talking to the people here, but remember harassment
[DISP-BibotII][BibotII]   will not be tolerated! Msg an op if you are being harassed
[DISP-BibotII][BibotII]   If you'd like to use my services, please
[DISP-BibotII][BibotII]   '/msg BibotII Hello' and I'll add you to the userlist!

Additionally, one can engage in one-to-one social control, either by employing the /ignore command (which renders private messages from specific individuals un-receivable) or through private channels as in the following example:

[DISP-playful][playful] hi..can you play today?
[DISP-playful][playful] I'm 25/6'/185/blond/hazel/broad shoulders/romantic/love to play on the phone...and you?
[DISP-playful]<Marcia_Br>   well, i'm blonde, too. but other than that, i don't think we have much in common. . .
[DISP-playful][playful] okay...*hangs his head* sorry to bother you
[DISP-playful]<Marcia_Br>   no problem
[DISP-playful][playful] hi..can you play tonight?
[DISP-playful]<Marcia_Br>   we've been through this before. go away.
[DISP-playful][playful] you got it...won't bother you again..have fun

A fair number of interactions which may be defined as unwanted messages and/or comments occur when individuals violate the distinctions between public and private channels; they attempt 'private type' discussions within the public format. In some cases, such as the one below, individuals in the channel drop messages that are intended for private channels:

[DISP-#35+home][Hodad]   ./msg TrishB You know I have always loved you deeply.

[DISP-#35+home][TrishB]   SHY? is that you?
[DISP-#35+home][SweetNes]   OH GOD....here we go
[DISP-#35+home][Hodad]   woops.
[DISP-#35+home][Nim]   THANK YOU TRISH XOXOXO
[DISP-#35+home][SweetNes]   Trish: yes
[DISP-#35+home][Nim]   LOL
[DISP-#35+home][TrishB]   *hugs* NIM
[DISP-#35+home][TrishB]   LOL

In such instances, the dropped messages are commented upon by other members in such a way that the perpetrators understand that they should be more careful. Contrast these mix-ups, defined as relatively minor, between pubic and private communications with the example below. Initially, the regulars on channel #25plus take cynnee to be a 'normal' Undernetter; one who does not pose any sort of threat to the stable interactions of the channel. It becomes increasingly clear, however, that cynnee may be a threat to group values. Only moments after cynnee leaves the channel do the regulars fully realize what has happened and that her behavior warranted a /kick. With this option unavailable, the group members proceed to ridicule her behavior with a game of intentionally dropped messages. In such a simulation, they construct her as deviant and themselves as ones who would never engage in such behavior:

[DISP-#25plus][SERVER]   cynnee!~dfoster@dial08.lloyd.com   has   joined this channel
[DISP-#25plus][blondee]   hi cynnee
[DISP-#25plus][cynnee]   hi blondee
[DISP-#25plus][cynnee]   any guys in here??
[DISP-#25plus][Wild]   i am
[DISP-#25plus][friend]   I was the last time I checked
[DISP-#25plus][friend]   :)
[DISP-#25plus][cynnee]   wild...how old are you
[DISP-#25plus][Wild]   cynnee i am 32
[DISP-#25plus][cynnee]   really??????????????
[DISP-#25plus][Wild]   really!!!!!!!!!!!!!
[DISP-#25plus][Therapist]   is that old cyn?
[DISP-#25plus][cynnee]   not at all...that is good

[DISP-#25plus][cynnee]   tired of talking to 20 year olds
[DISP-#25plus][Therapist] pssstt cynnee... he is a nice guy...
[DISP-#25plus][friend] I was sweating that one wild
[DISP-#25plus][friend] heheheh
[DISP-#25plus][Therapist] heehee
[DISP-#25plus]<Marcia_Br>   lol
[DISP-#25plus][Wild]   hehehe
[DISP-#25plus][cynnee]   rats
[DISP-#25plus][Wild]   rats?
[DISP-#25plus][cynnee]   (wantin a rough rugged type)
[DISP-#25plus][Therapist] hmmmm
[DISP-#25plus][friend] hmmmmmm
[DISP-#25plus][Therapist] \msg cyn he sounds pretty good on paper.. I mean screen...
[DISP-#25plus][Wild]   hehehe
[DISP-#25plus][cynnee]   wild...what makes you so...wild
[DISP-#25plus][Wild]   cynnee i am what i am
[DISP-#25plus][cynnee]   is there a way I can page someone on here to see if they are near
[DISP-#25plus][blondee]   isn't that what the burning bush said to Moses, Wild?
[DISP-#25plus][Wild]   hehehe
[DISP-#25plus][cynnee]   lol
[DISP-#25plus][cynnee]   my bush is burning
[DISP-#25plus][Therapist] oh my
[DISP-#25plus]<Marcia_Br>   ??
[DISP-#25plus][ACTION]   Therapist covers her eyes
[DISP-#25plus][blondee]   ! :O
[DISP-#25plus]<Marcia_Br>   did i miss something...
[DISP-#25plus][Wild]   ehehe
[DISP-#25plus][cynnee]   thanks
[DISP-#25plus][CatTail]   I was not watching...what did you do wild...become single??
[DISP-#25plus][Wild]   hehehe
[DISP-#25plus][cynnee]   wild...for someone who is wild...he he he doesn't fit you
[DISP-#25plus][cynnee]   ciao room

[DISP-#25plus][blondee]   bye cynnee!
[DISP-#25plus][SERVER]   cynnee has left this channel
[DISP-#25plus][CatTail]   bee bii cynnee
[DISP-#25plus][CatTail]   aw...what did you say to her wild?
[DISP-#25plus][blondee]   was she upset?
[DISP-#25plus][Wild]   nothing i swear!
[DISP-#25plus][CatTail]   I do not know
[DISP-#25plus][Therapist]   \msg cynnee ok ... just type /join #netsex... you'll like it there better.
[DISP-#25plus][Therapist]   heehee
[DISP-#25plus][blondee]   Ther, LOL!
[DISP-#25plus][Therapist]   <---- bad girl
[DISP-#25plus]<Marcia_Br>   lol
[DISP-#25plus][blondee]   did she really say her bush was burning?
[DISP-#25plus]<Marcia_Br>   yep.
[DISP-#25plus][Therapist]   heehee
[DISP-#25plus][blondee]   good grief!
[DISP-#25plus][ACTION]   blondee is laughing
[DISP-#25plus][Wild]   / msg therapist spanking
[DISP-#25plus][Therapist]   oh my!
[DISP-#25plus][Therapist]   \msg wild ok... later..
[DISP-#25plus][CatTail]   ooops you guys are not connecting well
[DISP-#25plus]<Marcia_Br>   oh, i don't know...
[DISP-#25plus][Wild]   burning bushes can be cooled...
[DISP-#25plus][CatTail]   giggle.......
[DISP-#25plus][Therapist]   how did this conversation get started?

In other instances, the style of private interaction on the public channel are even more explicit. In this case, the perpetrator is quickly kicked from the channel:

[DISP-#chatzone][slugger]   lady I love you
[DISP-#chatzone][dastompie]   inky, i happened to be a beautiful black young lady thank you, how about you are you a girl or a boy
[DISP-#chatzone][floosie]   Scottish accents are the best in the world
[DISP-#chatzone][Lady]   floosie: I love ladies and men at the same time
[DISP-#chatzone][leifman]   lady take it to #netsex

[DISP-#chatzone][Emily2]   Wendy!!!!
[DISP-#chatzone][Lady]   slugger: well come and get it
[DISP-#chatzone][vez]   get on your own channel, baby.
[DISP-#chatzone][Wendy]   EM2!!!!!!!!!!!!!!!!!!!!!!!!!!!!!!!!!!
[DISP-#chatzone][floosie] bye then lady cause i don't
[DISP-#chatzone][inky]   i'm a girl..
[DISP-#chatzone][Lady]   leifman: you want me don't you
[DISP-#chatzone][Weasle]   floosie: why's that then?
[DISP-#chatzone][slugger] why don't you just give it to me, lady?
[DISP-#chatzone][leifman-KICK]   has kicked Lady from #chatzone have a nice day
[DISP-#chatzone][Emily2]   good one leif

In this instance, several individuals (including lady, floosie, and slugger) are implicated in the deviant activity, although lady is defined as the instigator. When leifman suggests that lady 'take it to #netsex', implying an inappropriate interaction in a 'respectable' public channel, she suggests that his problem perhaps is that he is 'left out' of the interaction. For this direct confrontation with an op, she is immediately kicked and leifman is subsequently congratulated by Emily2, another op in the channel. Floosie and slugger were kicked shortly thereafter.

Another type of behavior which can be defined as general harassment is 'insulting' a channel, or the Undernet, and by implication, all of the identities associated with it. In this example, an individual parodies the channel by changing his nick to _41plus, and then to 'suck'. He is subsequently splatter kicked from the channel and banned:

DISP-#41plus][SERVER]   _41plus!vfuoco@enigma.idirect.com has joined this channel
[DISP-#41plus][rutabaga] _41plus is cruising?
[DISP-#41plus][_41plus]   Kris: why'd you kick me off.
[DISP-#41plus][Kristyne] 41plus it was a mistake!
[DISP-#41plus][SERVER]   _41plus's nickname is now Suck
[DISP-#41plus][fleep]   and maybe it wasn't
[DISP-#41plus][kojac]   bye suck hehehhe
[DISP-#41plus][Heatgain] suck?...very mature..
[DISP-#41plus][Kristyne] No it wasn't

[DISP-#41plus][ACTION]   Heatgain pulls out a dentist's drill.
[DISP-#41plus][ACTION]   Heatgain straps suck into a dentist's chair.
[DISP-#41plus][ACTION]   Heatgain asks suck..."Is it safe?"
[DISP-#41plus][Heatgain-KICK]   has kicked Suck from #41plus ^ l PhoEniX l^
[DISP-#41plus][Cadd]   LOL
[DISP-#41plus][Kristyne-MODE]   Has changed Suck!*vfuoco@enigma.idirect.com's mode to +b
[DISP-#41plus][rutabaga]   Not a mistake, Kris!
[DISP-#41plus][mermaid]   LOL Heat!!!!
[DISP-#41plus][Cadd]   LOL
[DISP-#41plus][ACTION]   PennyLane watches Suck sail overhead
[DISP-#41plus][Dusti]   nice kick and ban ya'll
[[DISP-#41plus][Cocoa]   suck sucks :)

It is not uncommon, depending upon the situation, for channel regulars to want to 'have a little fun' with the deviant before kicking (and possibly banning) him or her. Such activity serves as a means through which channel regulars can reassert their status as 'member'. By openly ridiculing the behavior of the deviant while he or she is still in the channel, regulars simultaneously maintain the values and cohesion of the channel and the identity of individuals as members.

In the example below, nastyboy enters the channel and attempts to 'fit in' by emulating the actions of one of the regulars. However, because of the immediate problems caused by his nick and the manner in which he attempts to engage the regulars, he succeeds only in helping them define him as 'immature'. This, combined with Gyrfalcon's comment that he has encountered nastyboy on previous occasions and tried to tell him to 'get a new gig', provides grounds for labeling nastyboy as a 'hopeless case'; one who has ignored or rejected previous opportunities to refrain from rule-violating behavior. [Pfuhl and Henry 1993] The core members of the channel 'gang up' on him and ridicule his actions and his typing extensively before Iris eventually splatter kicks him:

[DISP-#41plus][SERVER]   nastyboy!^wldsaclin@saclink1.csus.edu   has joined this channel
[DISP-#41plus][nastyboy]   hi all!

[DISP-#41plus][miranda]   er...nastyboy!
[DISP-#41plus][TC]   nice kick Gyr!!...65 yards easy!!!!
[DISP-#41plus][ACTION]   nastyboy keeps quiet since there is kicking going on...heheh
[DISP-#41plus][ie] iris rofl
[DISP-#41plus][miranda]   gyr..makes ya tingle doesn't it
[DISP-#41plus]<Marcia_Br>   hi, acftdoc
[DISP-#41plus][AcftDoc]   Goooooooooooooooooooooood Afternoon!!!!
[DISP-#41plus][Da_Bear]   doc!
[DISP-#41plus][miranda]   hey jetdoc
[DISP-#41plus][AcftDoc]   Hello Marcia :-)** @-}--------------------
[DISP-#41plus][AcftDoc]   Hi Miranda :-)**
[DISP-#41plus]<Marcia_Br>   long time no see:)
[DISP-#41plus][ACTION]   nastyboy whistles for the truck to back up and dump the truck load of roses for the women here...
[DISP-#41plus][nastyboy]   @}--}----
[DISP-#41plus][AcftDoc]   Yes, it has...
[DISP-#41plus][miranda]   nasty...our men like roses too
[DISP-#41plus][Gyrfalcon] OH boy...
[DISP-#41plus][miranda]   LOTS of them
[DISP-#41plus][Cadd]   <====likes em
[DISP-#41plus][Iris]   OH..my favorite thing..CYBERROSES....sooooo sincere!!!!
[DISP-#41plus][miranda]   iris, don't you get that special feeling when DUMPED on by a TRUCK of roses?
[DISP-#41plus][sloe]   /m nasty ..you'd do far better with cyberco*k
[DISP-#41plus]<Marcia_Br>   lol
[DISP-#41plus][sloe]   oops
[DISP-#41plus][Iris]   Mir...sets me atingle!!!!
[DISP-#41plus][miranda]   sloe! LOL
[DISP-#41plus][Iris]   LOL SLOE!!!!!
[DISP-#41plus][ACTION]   Gyrfalcon has tried to tell nastyboy to get a new gig....before....<shrug>
[DISP-#41plus][TC]   sloe: was that cork??
[DISP-#41plus][Chantilly] sloe..only if he co*ks Italian
[DISP-#41plus][ACTION]   AcftDoc covers Marcia with cyber roses
[DISP-#41plus][Cadd]   LOLOL

Deviant Behavior: Its Definition and Control

[DISP-#41plus][miranda]   gyr...nice thought at least
[DISP-#41plus][nastyboy]   miranda I guess we are sweat smelling with thorns butthorns can always be cut to the hand of a woman...
[DISP-#41plus][miranda]   um..who would like the pleasure
[DISP-#41plus][Iris]   OH NASTY....that poetry is magic!!!!!!
[DISP-#41plus][Cadd]   what the f*ck does that mean???
[DISP-#41plus][Chantilly]   oh pass the waste basket..QUICK!
[DISP-#41plus][miranda]   ?
[DISP-#41plus][Gyrfalcon]   sweat smelling?
[DISP-#41plus][Da_Bear]   huh?
[DISP-#41plus][miranda]   anytime now...
[DISP-#41plus][ACTION]   Iris grabs nastyboy by the neck and props nastyboy's mouth open.
[DISP-#41plus][ACTION]   Iris stuffs large quantities of frozen ice cream sandwiches down nastyboy's throat.
[DISP-#41plus][ACTION]   Iris sets nastyboy off to the side to admire his new full size PEZ dispenser...
[DISP-#41plus][Iris-KICK] has kicked nastyboy from #41plus ^ I PhoEniX I^
[DISP-#41plus][Gyrfalcon]   is that like when you work out?
[DISP-#41plus][ie]   miranda rofl
[DISP-#41plus][miranda]   the sooner the better
[DISP-#41plus][miranda]   thanks iris:)
[DISP-#41plus][ACTION]   sloe vomits ....Next case!
[DISP-#41plus][TC]   LOL
[DISP-#41plus][fw]   bye nasty!
[DISP-#41plus][Cadd]   LOLOL
[DISP-#41plus][miranda]   ROFL
[DISP-#41plus][ie]   thorns butthorns
[DISP-#41plus]<Marcia_Br>   lolol
[DISP-#41plus][Iris]   LOL!!!!!
[DISP-#41plus][sloe]   hahahahahah
[DISP-#41plus][Gyrfalcon]   sloe: <g>
[DISP-#41plus][Iris]   What else goes with Sweat???
[DISP-#41plus][sloe]   its gunna be a good day after all

It is clear from this interaction that nastyboy has been in this channel before and has refused to change his nick to one which is more appropriate for a crowd of people who are at least 40 years old. However, the outcome may still have been different for nastyboy had he not directly challenged the authority of miranda, who is a regular and a member of the 'core' in this channel. Once this mistake is made, recovery is impossible; everything nastyboy says is ridiculed without mercy even after he is splatter kicked from the channel.

And finally, one can occasionally encounter a case of deviance in which the perpetrator is defined as so threatening that an orchestrated vindication, not unlike a criminal trial and sentencing, is deemed warranted by the channel members. In this example, the individual in question is implicated in all of the forms of deviant behavior described above. He has a problematic nick, he has intentionally and repeatedly engaged in harassment through private messaging (which included profanity and slander with regard to personal identity), and has refused all attempts at reason. Under these circumstances, the individuals involved chose to deal with him collectively, rather than individually, and the style of interaction eventually evolved into a collective behavior in the form of a mob:

[DISP-Jeannette]<Marcia_Br>   by the way, what did bigpete say/do to deserve this?
[DISP-Jeannette][Jeannette]   He has been mercilessly messaging us all, all day long without stop
[DISP-Jeannette]<Marcia_Br>   each woman individually, or the channel?
[DISP-Jeannette][Jeannette]   Each and every one individually
[DISP-Jeannette]<Marcia_Br>   Good grief. he must have a lot of free time...
[DISP-Jeannette][Jeannette]   Yes indeed

[DISP-#lesbian][annalis]   ok, let's get him frustrated!
[DISP-#lesbian][mel]   oh..all right
[DISP-#lesbian][Elizabeth]   same here...I can go for anything so long as no one gets hurt..
[DISP-#lesbian][Jeannette]   OK BigPete is unbanned and un kicked :)
[DISP-#lesbian][mel]   cool
[DISP-#lesbian][mel]   someone invite him
[DISP-#lesbian]<Marcia_Br>   who is bigpete?

[DISP-#lesbian][mel]   marcia: an arrogant male chauvinist..we want to put him in his place
[DISP-#lesbian]<Marcia_Br>   but, why did you unban him then?
[DISP-#lesbian][Aeryn]   Marcia: I think the women here want to give him a *personal* treatment...
[DISP-#lesbian][ACTION]   annalis can't help laughing... oh no!!
[DISP-#lesbian][ACTION]   mel pulls out the meat tenderizer *giggle*
[DISP-#lesbian][ACTION]   Aeryn dares someone to change nicks to Dr.Bobbit...;)
[DISP-#lesbian][mel]   haha
[DISP-#lesbian][SERVER]   bigpete!nupp@128.250.50.72 has joined this channel
[DISP-#lesbian][bigpete]   cool
[DISP-#lesbian][bigpete]   annalis: wanna have sex?
[DISP-#lesbian][Jeannette]   Lorena was stupid Aeryn! Criminally stupid!
[DISP-#lesbian][mel]   ARG!!!
[DISP-#lesbian][annalis]   Peter, Peter... you stupid wiener! had a woman but couldn't keep her!!
[DISP-#lesbian][Jeannette]   I mean, SHE HAD A GARBAGE DISPOSAL!!!
[DISP-#lesbian][mel]   bigpete: you are a chauvinist pig and lesbian hater!!
[DISP-#lesbian][bigpete]   hey!!
[DISP-#lesbian][bigpete]   i'm good in bed!
[DISP-#lesbian][bigpete]   i am not mel whoever you are!
[DISP-#lesbian][Elizabeth]   really all those who claim to be great usually suck big time!
[DISP-#lesbian][annalis]   prove it bigpete..
[DISP-#lesbian][mel]   you pull you dick all the time don't you
[DISP-#lesbian][bigpete]   i don't
[DISP-#lesbian][Elizabeth]   all you do is play with yourself because you can't get anything!
[DISP-#lesbian][bigpete]   i can't prove it
[DISP-#lesbian][ACTION]   LorenaB pulls out the scissors!
[DISP-#lesbian][Jeannette]   I have unbanned him and taken him from the kick list
[DISP-#lesbian][bigpete]   but i don't all the time elizabeth
[DISP-#lesbian][ACTION]   LorenaB moves towards bigpete
[DISP-#lesbian][bigpete]   but what are you doing

[DISP-#lesbian][bigpete]   please be nice
[DISP-#lesbian][ACTION]   Bobbit pulls out a scalpel
[DISP-#lesbian][ACTION]   LorenaB raises the scissors!
[DISP-#lesbian][ACTION]   Aeryn gets the crowd aroused chanting, "SNIP, SNIP, SNIP...etc."
[DISP-#lesbian][ACTION]   Bobbit approaches BigPete cautiously
[DISP-#lesbian][ACTION]   mel grabs bigpete by his little balls and squeezes
[DISP-#lesbian][bigpete]   hey!!!
[DISP-#lesbian][bigpete]   not fair
[DISP-#lesbian][annalis]   you should like that!
[DISP-#lesbian][bigpete]   your ganging up on me
[DISP-#lesbian][ACTION]   Bobbit takes a swipe at his puny balls..
[DISP-#lesbian][annalis]   isn't that what you want?
[DISP-#lesbian][Bobbit]   Slice him..Dice him,...
[DISP-#lesbian][Bobbit]   castration!!!!
[DISP-#lesbian][ACTION]   Jeannette winds some razor wire around a broomstick and spreads Pete's legs
[DISP-#lesbian][ACTION]   mel crushes what's left of his wretched testicles
[DISP-#lesbian][bigpete]   but not like that
[DISP-#lesbian][ACTION]   Bobbit holds him down..give it to him..
[DISP-#lesbian][bigpete]   that hurts me stop it
[DISP-#lesbian][Bobbit]   let him know what its like..
[DISP-#lesbian][ACTION]   Jeannette applies some Ben Gay to Pete's inviting ass hole
[DISP-#lesbian][bigpete]   no don't...please be nice..i'm just a little man
[DISP-#lesbian][Bobbit]   ouch ...that must feel nice huh?
[DISP-#lesbian][ACTION]   Jeannette shoves the razor wired broom up Pete's ass and twists it a few times
[DISP-#lesbian][mel]   you were quick to abuse us before weren't you!
[DISP-#lesbian][ACTION]   Bobbit sits on his face and pees..
[DISP-#lesbian][Aeryn]   bigpete: we *knew* you were a little man or you wouldn't have needed that nick.
[DISP-#lesbian][bigpete]   yuk..that's sick bobbit
[DISP-#lesbian][ACTION]   annalis gets out the magnifying glass..
[DISP-#lesbian][Bobbit]   sick treats for sick men..
[DISP-#lesbian][bigpete]   yuk you're all nasty to me...you dyke woman
[DISP-#lesbian][Bobbit]   make sure you focus the sunlight on it..

[DISP-#lesbian][mel]   you are truly sick pete
[DISP-#lesbian][annalis]   I told you to be NICE!!!!
[DISP-#lesbian][ACTION]   Jeannette begins to fuck Pete up the ass with the razor wired broom taking nice long strokes
[DISP-#lesbian][annalis]   apologize NOW!
[DISP-#lesbian][Bobbit]   burn the little thing to a crisp..
[DISP-#lesbian][mel]   annalis: kick his balls
[DISP-#lesbian][ACTION]   Bobbit winds up and Pow!
[DISP-#lesbian][bigpete]   anna: but there ganging up on me...i just wanted to have some fun
[DISP-#lesbian][annalis]   do it bigpete.. or we will ignore you forever :)
[DISP-#lesbian][Bobbit]   say you're sorry!
[DISP-#lesbian][bigpete]   but
[DISP-#lesbian][Bobbit]   or I'll kick you again..
[DISP-#lesbian][annalis]   you can always choose to leave..
[DISP-#lesbian][bigpete]   ana: i can't
[DISP-#lesbian][bigpete]   i'm sorry...
[DISP-#lesbian][Elizabeth]   LOUDER!
[DISP-#lesbian][Jeannette]   So Mr Big Pete do you understand just how intrusive and disgusting we find you?
[DISP-#lesbian][ACTION]   mel gives pete's little penis a twist
[DISP-#lesbian][bigpete]   why are you hurting me...i wouldn't hurt you all at once
[DISP-#lesbian][ACTION]   Elizabeth watches as his penis crumbles to the ground as dust..
[DISP-#lesbian][bigpete]   but i like my dick
[DISP-#lesbian][annalis]   because you don't treat other people with respect!!!
[DISP-#lesbian][Elizabeth]   looks like the magnifying glass did the trick..
[DISP-#lesbian][mel]   all at once!!!?!?!?
[DISP-#lesbian][annalis]   shrivvvvvvvel..
[DISP-#lesbian][Esta]   Jeanette: I see we are sorting out, the odd intruder!
[DISP-#lesbian][Jeannette]   Yes indeed Esta, taking out a rather lame man bent on net sex
[DISP-#lesbian][Aeryn]   esta: You bet. :) I dunno what he did, but he got everyone mad enough to do *this* to him...:)

[DISP-#lesbian][Esta] BigPete - small dick: I would say you are not welcome!, just outsider observation!
[DISP-#lesbian][bigpete]   you are only women
[DISP-#lesbian][annalis]   AHHHHH.. ok, that's it...
[DISP-#lesbian][mel]   ARGGG!!!!
[DISP-#lesbian][annalis]   you pissed me off.. and that's impossible to do!
[DISP-#lesbian][ACTION] Aeryn says that now that we have his dick taken care of perhaps we should deal with other parts...
[DISP-#lesbian][ACTION] mel kicks pete in the head hard!
[DISP-#lesbian][Jeannette] Aeryn aw c'mon we are just having fun with him.... like his little fun messages were to us :)
[DISP-#lesbian][Elizabeth] How about it gals!
[DISP-#lesbian][Elizabeth] Yeah don't you just love it!
[DISP-#lesbian][Jeannette] Goddess! I wish I had IRC logged this one!
[DISP-#lesbian][SERVER] allissa!TWWcrl.co@obelix.wu-wien.ac.at has joined this channel
[DISP-#lesbian][Esta]   Aeryn: Good to know, I wonder why he is still here, groveling!?
[[DISP-#lesbian][ACTION]   mel spreads pete's legs again..and offers everyone a free kick, knee or grab *giggle*
[DISP-#lesbian][ACTION] Jeannette removes the broom *fast*
[DISP-#lesbian][bigpete]   your are bitches all of you woman
[DISP-#lesbian][Jeannette] and offers the bloody device to Pete to suck on now
[DISP-#lesbian][Elizabeth] Go get laid for real kid..then talk to us,ok?
[DISP-#lesbian][ACTION] annalis nods!
[DISP-#lesbian][Elizabeth] Eat it bastard!
[DISP-#lesbian][Esta]   bigpete: That was stupid, go jerk one of your friends off on another channel!
[DISP-#lesbian][bigpete]   you're all dirty...i am a man!
[DISP-#lesbian][ACTION] Jeannette stuffs the razorwired broom down Pete's throat
[DISP-#lesbian][ACTION] Aeryn rouses the other bloodthirsty monsters with a cry of "Brains!" and then stops realizing that the prospective prey has none...
[DISP-#lesbian][Jeannette] Shall we remove this piece of toxic waste now?
[DISP-#lesbian][bigpete]   why do you like hurting guys balls?

[DISP-#lesbian][ACTION] Aeryn thinks he *liked* Jeannette's broomstick treatment in his heart of hearts but would *never* admit it...
[DISP-#lesbian][bigpete]   my balls? i am a man
[DISP-#lesbian][robynn]   jeanette like the ad sez - JUST DO IT
[DISP-#lesbian][Elizabeth]   because it HURTS thats why!
[DISP-#lesbian][bigpete]   i am not a poofta
[DISP-#lesbian][robynn]   areyn, agreed
[DISP-#lesbian][annalis-MODE]   Has changed *!*@128.250.50.*'s mode to +b
[DISP-#lesbian][Jeannette] Spinning the Wheel of Fate... It lands on Splatterkick 30!
[DISP-#lesbian][ACTION] Jeannette dumps a can of gasoline on bigpete
[DISP-#lesbian][ACTION] Jeannette lights a match and tosses it.
[DISP-#lesbian][ACTION] Jeannette laughs as bigpete turns squirms and smokes to a blackened crisp.
[DISP-#lesbian][Jeannette-KICK]   has kicked bigpete from #lesbian ^ I PhoEniX I^

[DISP-#lesbian][Jeannette] There!
[DISP-#lesbian][mel]   haha...is he gone
[DISP-#lesbian][Jeannette] We all got to show him *just how much* we liked him and his messages
[DISP-#lesbian][robynn]   nicely done but i'll bet two C batteries that he will be back
[DISP-#lesbian][ACTION] Aeryn doesn't think he'll show his lame-o self here again, that's for sure.
[DISP-#lesbian][Elizabeth] Who knows he might like what we did to him..the attention and all..
[DISP-#lesbian][ACTION] Aeryn suggests synchronizing /ignore lists...
[[DISP-#lesbian][robynn]   aeryn me thinks he likes/needs abuse
[DISP-#lesbian][Elizabeth] sick soul..)=
[DISP-#lesbian][ACTION] Elizabeth frowns in disgust..
[DISP-#lesbian][Aeryn]   robynn: probably more like he needed the attention, and personally I think we fed it to him. The worst thing we could have done to him is ignore him.
[DISP-#lesbian][Aeryn]   It's funny...the only time he broke off from his anti-woman, anti-les spiel was to deny he was queer.

[DISP-#lesbian][robynn]   aeryn, sadly i think your right, we also supported his deep misogyny and fear of women
[DISP-#lesbian][ACTION] mel wonders what we would do with someone like him if this was a real-life room and he came into it in real life *giggle*
[DISP-#lesbian][ACTION] Elizabeth laughs and chuckles.
[DISP-#lesbian][Aeryn]   robynn: actually, i think he was genuinely surprised by the level of animosity in the room...didn't think us capable of it.
[DISP-#lesbian][mel]   robyn: yup

Clearly, participants in the Undernet see themselves as belonging to a community that is based upon the basic values of friendliness and cooperation. Rules of behavior have been constructed that are intended to allow all members of the community to participate in the interaction, provided that they agree to follow those rules. That these individuals see themselves as members of something larger than themselves is evidenced by the their willingness to punish rule violators in an effort to preserve the integrity of the group.

Not only are channel members willing to punish rule violators, they have developed elaborate systems of classifications of deviant behavior and various means for deciding what sorts of punishments are warranted. While there exist some basic rules of behavior, it is understood by the members of the Undernet that each definition of deviance is highly situational. Allowances are made for unintended violations of social norms, provided that the individual in questions furnishes the universally accepted account of such behavior in the form of an excuse [Scott and Lyman 1975]: 'I'm sorry, I'm a newbie'. Community members utilize the frameworks of both public and private channels to define someone as deviant or non-deviant. It is through such mutual aligning of thoughts that deviance can be said to be constructed in interaction.

Beyond being willing to punish transgressors and capable of defining transgressions in fine gradations, Undernetters have access to social control mechanisms ranging from the joke, to sarcasm and ridicule, to the /kick and /ban commands. Depending on how the situation is defined by those involved, these methods may be employed in this order or a channel op may immediately opt for the /kick and /ban commands. The point is that the channel is seen as a community by its regulars; something worth fighting for and maintaining.

THE USENET COMMUNITY

The central concerns in the Usenet community are the maintenance of the technical efficiency of the network and maintenance of respect for community values. All Usenet norms are developed in order to perpetuate the community and its individual newsgroups by defining the purposes of Usenet and the types of interactions that are appropriate or inappropriate. Technological distinctions between Usenet and the Undernet make it possible for an individual in one newsgroup to affect the entire Usenet system, while the Undernetter can only affect one channel at a time. Therefore, when thinking about deviant behavior in Usenet, one must understand the distinction between those behaviors which affect the community as a whole and those which only affect one (or a few) newsgroup(s).

The most commonly occurring and potentially threatening type of deviant behavior that affects the community as a whole, which the entire community defines as deviant, and which requires a community-wide response, is the activity of 'spamming'. According to the Usenet Site Administrator's Guide to Netiquette [1994], there is a general community-wide consensus that certain activities constitute net-abuse, the most widely known being spamming:

> A user or group of users is said to have spammed when he/she/they posts one or more messages with substantially the same content to a large number of newsgroups, in many of the which the post is off-topic. A spamming infraction is more serious if the spam is not cross-posted to the newsgroups involved, but is instead posted to each group separately. This means that the newsreader software at the receiving end will not be able to determine that the message has already been read in another newsgroup, and the receiving users will be repeatedly confronted with the post. Spamming also wastes bandwidth and disk space, and is especially annoying to sites with slow communication links and limited disk storage. A post or posts need not happen all at once to be considered spam. Usually spamming occurs when a user attempts to use Usenet as a one-way broadcast medium, instead of as a forum for discussion and exchange of ideas. Much of the poor press coverage of Usenet comes about because of a lack of understanding of this distinction. Usenet is not intended as a means of getting people who are not already interested in a topic to

pay attention to it. There is no hard and fast definition of "spamming". However, Usenet groups are sufficiently distinct that it is quite rare that an article is really on-topic in more than ten or so groups. [Usenet Site Administrator's Guide to Netiquette]

Thus, even the definition of spam is highly contextual within Usenet. And while spam is generally thought of (in the popular press) as advertisements, this is not necessarily the case. Properly tailored posts, even when they are of a marketing nature, do not necessarily violate the norms of Usenet. What is important in the concept of spam is the number of duplicate postings that the individual makes and to how many newsgroups. And again, unlike the depictions in the popular press, defining what constitutes spam and how it should be dealt with are structured, community-wide processes.

According to the news.admin.net-abuse FAQ [1995] and the Usenet Site Administrator's Guide to Netiquette [1994], the organized defining and elimination of spam began after a particularly threatening round of spamming now known as the 'Green Card' spam of April, 1994, during which "...an ad appeared in virtually every Usenet group. The perpetrators announced their intention to spam repeatedly, and the Usenet access provider did not take quick action to prevent them from doing so. Their actions threatened Usenet as a whole. In this circumstance, a programmer wrote a 'cancel-bot'[2] which forged a cancellation message for every post originating from the spammer's account. When the 'Green Card' ads appeared again, the cancel-bot prevented their propagation. This cancel-bot was generally applauded throughout the Usenet community."

The individual who wrote the spam cancelling program came to be known as the Cancelmoose [tm], and, according to the news.admin.net-abuse FAQ [1995], is thought of as:

"...the greatest public servant the net has seen in quite some time. He or she sends out spam-cancels and then posts notices anonymously to news.admin.policy, news.admin.misc, and a.c-e.n-a (alt.current-events.net-abuse). The Moose stepped to the fore on its own initiative, at a time when spam-cancels were irregular and disorganized, and has behaved altogether admirably-- fair, even-handed, and quick to respond to comments and criticism, all without self-aggrandizement or martyrdom. Cancelmoose[tm] appears to have near-unanimous support from the readership of all three above-mentioned groups" [p. 8]

When the Cancelmoose[tm] created its cancel-bot, it also created several institutional strategies for legitimating the definition and control of deviance, including the 'cancel' and 'cyberspam' conventions. In order to legitimate the cancelling of spam, the canceler must confirm the spam him- or herself; one cannot simply take the word of other Usenetters that someone is spamming newsgroups. Additionally, the canceler must announce his or her actions to the news.admin.net-abuse newsgroup in order to prevent confusion and assure Usenetters that something is being done to clear up the problem. Finally, the canceler must employ the 'cancel' and 'cyberspam' conventions for cancelling spam, where the 'cancel' convention allows one to create a cancel message ID by adding 'cancel' to the original ID, and the 'cyberspam' convention allows sites that do not wish to accept the cancels to opt out of the process. Usenetters may locate announcements of spams and spam cancels in the news.admin.net-abuse.announce newsgroup.

Community members define spam as a real threat to the technical and social integrity of the community as a whole. Spam left unchecked can overload entire Usenet sites, particularly the small ones, and prevent the timely propagation of relevant posts. As the Usenet system grew and became more well known to individuals outside of the computer programmer subculture from which it emerged, the incidence of spam continued to increase. Usenetters defined the 'Green Card' spam incident as the proverbial last straw--the point at which something had to be done.

However, even though spam is a threat to the network, Usenetters still attempt to treat spammers with leniency. This is similar to attempts made by Undernetters to give individuals who may not have known any better (the 'true' newbie) the benefit of the doubt; someone unfamiliar with the norms of Usenet may recover from an inadvertent spamming through a public apology in the news.admin.net-abuse.misc newsgroup. In fact, given the potential that spammers have to destroy the technical and social integrity of Usenet, the 'punishment' for such behavior is relatively slight.

In addition to spam, other forms of net-abuse include 'flooding' newsgroups and follow-ups to misc.tests, sendsys bombs and other mailbomb posts. Similar to Undernet, 'flooding' in Usenet occurs when ". . .a user or group of users posts so many messages to a group that it is rendered unusable.

Posts may be on-topic or not, but if the action prevents other users from exchanging ideas through that newsgroup, it is considered net-abuse." [Usenet Site Administrator's Guide to Netiquette]

One can imagine how disruptive this particular sort of behavior would be to the stability of ongoing conversations in the newsgroup. If someone chooses to replicate his or her post say, 1000 times, and flood a newsgroup, this can potentially clog small Usenet sites. And of course, the problem is magnified if that article is cross-posted to more than one group. Paying Usenetters then log in to their regular groups only to find that they have 1000 copies of the same post, and no other articles, since one's newsgroup mailbox will only hold so many articles at a time.

These examples do not exhaust the list of net-abuse activities that individuals may engage in; spamming, flooding and mailbombs represent those activities which are generally accepted as net-abuse by the community as a whole. In order for Usenetters to decide what sorts of activities constitute deliberate net-abuse, they have created a newsgroup hierarchy, news.admin.net-abuse, to debate which types of activities warrant being defined as sufficiently deviant. Within the news.admin.net-abuse hierarchy, there are two groups, news.admin.net-abuse.misc and news.admin.net-abuse.announce. In the former group, individuals debate whether or not certain actions constitute net-abuse and warrant a community-wide response. In the later group, spam and spam cancels (and other forms of deviant behavior) are announced to the general Usenet public. [News.admin.net-abuse FAQ]

It is not the responsibility of 'common' Usenetters alone to defend the Usenet community against threats such as these. Sysops, through coordinated efforts, also play a central role. According to the Usenet Site Administrator's Guide to Netiquette:

> Once you become a sysadmin, the rest of the Usenet community will expect that you are prepared to discipline your users when they engage in net-abuse. . . In general, it is best if posters resolve complaints among each other, without system administrators getting involved. However, in many cases it is obvious that what is being complained about really is intentional net-abuse, and in that case you should act immediately.

If someone complains to you about some posts, and you agree that the posts in question are really net-abuse, then you should send out cancels for the offending articles, and take action to prevent repetition of the act. If you don't believe that the questionable posts are net-abuse, you should send reply e-mail to the complainer(s) explaining why. Any site whose administrator doesn't make a good-faith effort to halt net-abuse originating at that site quickly gets labeled a renegade. A renegade site frequently becomes the target of a boycott. This means that other sites will refuse to feed it articles, and will refuse to carry any of that site's posts. This situation is not too common.

If an abuse is reported to you, you have pretty broad discretion as to how to handle it. However, other site administrators will be quite angry if it doesn't stop quickly. You may have to deny net-abusers Usenet access if all else fails. Withdrawal of Usenet privileges may take place even if other forms of Internet access remain. . .You definitely can't allow net-abuse . . . How you react will be up to you. Of course you want your site to retain good relations with the rest of the Usenet community.

Besides behaviors which have been classified as net-abuse by the Usenet community as a whole, there are a variety of within-group behaviors that may also be defined as inappropriate, but which require only a within-group response in return. Some of these behaviors are group-specific parallels of community-wide net-abuse, and some are behaviors which are related to the identities of the groups (and group regulars) themselves. In their exploratory study of five newsgroups, McLaughlin, Osborne and Smith [1995] develop an excellent 'taxonomy of reproachable conduct', in which they identify seven basic types of deviant behavior: incorrect/novice use of technology; bandwidth waste; violation of network conventions; violation of newsgroup-specific conventions; ethical violations; inappropriate language; and factual errors. [pp. 97-98]

What follows are examples of some of these deviant behaviors and an explanation of how the context of those behaviors determines the types and severity of social control methods that Usenetters employ to reestablish social order. The first two categorizations of deviant behavior, incorrect or novice use of technology and bandwidth waste are those which parallel the community-wide net-abuse issues discussed above. Just as spamming, flooding and mailbombs use up bandwidth and waste the time of Usenetters across the net, empty messages, inappropriate reply to groups, posting multiple copies of same article, excessively long sigs, inclusion of excessive information in the sig, and so forth waste the time and bandwidth of members of individual newsgroups. As the name suggests, deviant behaviors classified as novice use of technology are general indicators that one is dealing with a newbie. Therefore, the means of dealing with such problems are usually mild. Likewise, many of the behaviors in category two are also indicative of newbie status, and warrant only mild responses from other group members.

'Reproachable conduct' categories three through seven, however, are ways in which newsgroup members may be classified as deviant because they represent some threat to either the group identity or the identities of particular individuals within the group. For example, if one posts a message with an inappropriate subject line, or changes the subject line in a thread for no apparent reason, the continuity of newsgroup threads, and therefore the identity of the group, is potentially lost. Similarly, if one 'creatively' edits the posts of another member, whether intentionally or not, one threatens the very identity of the individual being quoted. One's entire identity in Usenet is accomplished and presented through posts; as a rule, a violation of norms which threatens group and/or individual identity is met with harsher response from group members.

The means of social control that Usenetters have developed to deal with these various forms of deviant behavior bear striking resemblance to those developed by Undernetters; though technical distinctions exist, the end results are largely the same. For example, just as Undernetters have the private channel and the /ignore command as one-to-one means of social control, Usenetters have private email and the kill file.

According to the news.admin.net-abuse FAQ [1995], private email should be used as the first line of defense in controlling deviant behavior. One is encouraged to use email to discuss potentially problematic situations in the hopes that some mutually acceptable definition may be reached. Email is the preferred method of accomplishing social control because newsgroup members can simultaneously solve behavioral problems while keeping the newsgroup threads free from discussions of problematic behavior--discussions which the deviant often winds up controlling. Further, the use of email allows for more discrete socialization of newbies, without subjecting them to unnecessary public humiliation. As Hardy [1993] notes, "Once a user has access to Usenet, the first and gentlest measure of social control is feedback through personal E-Mail. If a new user ("clueless newbie") gets out of line, someone might be nice enough to cue them in privately. . ." [p. 8]

The kill file is another means of social control that Usenetters have at their disposal. And, like the /ignore command in IRC, individual group members may 'synchronize' the use of the kill file so that an intruder is effectively eliminated from posting to the newsgroup. That is, if one individual finds someone's actions in the newsgroup objectionable, that individual may put the posts of the person in question into a kill file, thereby creating a kind of customized personal boundary. When the objectionable individual posts to that group (or to any other) the person's newsreader will simply filter out those posts, and the newsgroup thread will appear as if those posts, and therefore that person, never existed. Like the /ignore command in IRC, the kill file reduces a member's reliance on the larger group's ability to define and defend a newsgroup's boundary. [Kollock and Smith 1994; MacKinnon 1995] To the extent that all members of a given group put the same individual in their personal kill files, the existence of that poster in the newsgroup is eliminated. Recall the example from Chapter Four, in which Farmer Fischer flames the rec.pets.cats newsgroup:
Farmer Fischer (dacktyl@winternet.com) wrote:
: Heather Thistle (hthistle@bbnplanet.com) wrote:

: : I wrote a few weeks ago about my mom's 8 year old female persian. She
: : was peeing all over the house.

: : Any and all info will be very much appreciated!

: An ice pick at the base of the skull jammed in real quick. No bleeding if you
: do it right. Then send the pussy to glory in a Gladbag[tm].

And Camille responded:

sigh

Please ignore Farmer Fischer. He's just some git from Minneapolis who has nothing better to do with his time than troll for flames by posting the most gross and rude crap he can think of to whatever newsgroup. Complaining to his ISP will do no good, as Winternet does believe in free speech and will not censor any users.

Please learn to use a kill file, and put the Farmer in it. Either that, or skip over his postings altogether if you do not agree with them. Thanks.

--Camille Klein, a Winternet user and no fan of Farmer Fischer.

In this example, Camille attempts to dissociate herself from Fischer by commenting that, while she may have the same Matrix access provider as he does, she by no means agrees with his posts. Further, she encourages that the upstanding members of rec.pets.cats 'synchronize' their kill files by putting Fischer in it, thus eliminating his presence from the group. It is interesting to note, however, that many of the group regulars have chosen not to do this. At least some of the regulars of the rec.pets.cats group continued to read Fischer's posts and respond to them. The implication of this sort of activity is that, like Undernetters, Usenetters allow a certain amount of deviant behavior within their newsgroups, because it provides them with a means of reasserting their values and their identities as members. To the extent that they have individuals available who can be readily defined as deviant, they can continue to define themselves as 'good guys'.

While Usenetters can use private email and the kill file to maintain social order, much of the work of social control is accomplished openly in the newsgroup threads themselves. Were it not, Usenetters would have no means by which to continually reestablish the boundaries of the group or their status as regulars. Like Undernetters, Usenetters judge whether or not inappropriate behavior was likely committed inadvertently by a newbie or purposely, by an experienced Usenetter. And the initial clues in this definition are again, whether the regulars recognize the individual in question (whether they have encountered him or her before) and whether the action in question is one which could have been accomplished by a newbie or whether it required the skills of a technically competent individual. Usenetters utilize the public forum of newsgroup threads to define the situations and to mete out the appropriate response. As Kollock and Smith [1994] note:

> A successful newsgroup depends on its members following rules of decorum. What counts as acceptable behavior can, of course, vary

tremendously from newsgroup to newsgroup . . .Whatever the local rules of decorum, it is important that most participants follow them. . . . What participants can do is use a variety of informal sanctions to try to shape behavior. (Those who violate norms) might be insulted, parodied, or simply informed that their actions are undesirable. Often the response is both intense and voluminous. . .some actions step clearly out of the bounds of acceptability. These kinds of informal social control mechanisms depend upon moral suasion to have an effect. . .many people report that informal sanctions do have a significant effect on their behavior. [pp. 13-14]

The following example illustrates the ways in which Usenetters call upon community wide norms and institutions to define within-newsgroup situations, thereby reproducing and legitimating Usenet values. Group members Paul, Dawn and John are engaged in a conversation about the extent to which certain threads should be cross-posted to various alt.support.* groups (a potential reproachable conduct category two violation of 'indiscriminate cross-posting'). All are experienced Usenetters, and all are willing to abide by whatever definition of 'excessive' cross-posting gets established for this particular situation. General Usenet norms of politeness and concern for the other, as well as a general acknowledgement of the rules about cross-posting are employed:

In Article <47ub2b$s8b@fnnews.fnal.gov>
dammern@fnalv1.fnal.gov (Dawn Marie Mueller a.k.a. DAMMERN x8131) writes:

Paul writes:

>>Folks, this thread is getting cross-posted to a LOT of newsgroups, including
>>alt.support.depression.manic, which I follow, but it seems to me no longer
>>to be centrally relevant to the group. Please trim the list of groups to which
>>you cross-post. Thanks.

>>Paul

< snipped >

>I take Xanax. I am concerned and interested but have noticed that I have
>been hopping on and off the med (and have experienced the withdrawal
>symptoms) over the past few weeks, in part due to this discussion. I have
>been and will continue to follow this on sci.med, bearing that in mind.
>Freedom of speech is good, but let's be sensitive. Does anyone besides Paul
>wish to limit the distribution?

> -- DMM

IMO only, any discussion which is cross-posted to so many newsgroups, *tends* to branch into many different paths. When it does, it can't possibly be relevant to all of newsgroups to which it is cross-posted.

Additionally, when it is cross-posted to *support* newsgroups, many of these different branches are not welcome. I would suppose *most* people who read *support* newsgroups are looking for *support*. If they desire stimulating discussions or scientific reference, they can find them in other newsgroups.

I for one, would prefer the list be trimmed.

Regards,

John Daly

    There are also instances in which individuals feel badly enough about having violated Usenet norms that they publicly admonish themselves without any prompting from others. This tends to be the case in groups which lie toward the problem-solving end of the continuum described in Chapter Five, mainly because so much of the highly personal self is invested in these groups that one is immediately and deeply aware that a breech of norms represents a real threat to group cohesiveness. Often in these instances, other members will respond with statements to the effect that they understand and forgive the transgression[3]

Subject: Public apology for inappropriate posts
Date: 22 Sep 1995 05:43:18 GMT
From: Melissa Porter <melissa@hurtwood.pdial.interpath.net>

Organization: None
Newsgroup: alt.support.depression

In the past few days, I have allowed my personal feelings to overcome my common sense and I've posted articles and responses here that I regret having ever written, much less made public.

I could cite plenty of reasons why I'm on edge right now, the most immediate being that I have not slept for going on 40 hours as I write this. But nothing that is going on in my personal life right now excuses my behavior in a public forum, or anywhere else for that matter.

I could cite everything I wish I hadn't posted, but I think it must be obvious to the most casual observer which ones I'm referring to.

The one exception to this avoidance of further public self-flagellation is tonight's movable flame war against Blue Moon. I still disagree with his proposal, but it was neither necessary nor constructive for me to follow him from post to post with a blowtorch. I allowed the content and tone of his posts to push a few buttons I thought I'd done away with. I naively thought I was all finished with outbursts of uncontrolled anger. I was wrong, and I'm sorry to have exposed such a large audience to it.

If I've hurt anyone else's feelings, I apologize for that too.

No, I'm not going to make another self-aggrandizing farewell speech. I'm just going to try harder to keep my mouth shut when I feel my mental temperature beginning to rise.

Whoever used to use the sig "post in haste, repent in leisure" knew what they were talking about.

Melissa.

Among the many sympathetic responses to Melissa's post were the following:

Melissa, GO TO BED, AND GET SOME SLEEP N O W ! ! !

Sorry for shouting, but (bearing in mind I'm seeing all this hours behind the rest of the world) I don't want you to harm yourself anymore.

Don't mail or post back. I'm shutting down *now*.

Just go to bed, get some sleep, and we'll discuss it later - ok?

GoodNight / GoodMorning

Luv & Hugs,

Martin

and:

Oh, *bullshit*, Melissa! Don't let that stupid, irrational guilt (which sounds rather similar to what I've been feeling all this week), tell you you did something wrong, at least in Blue's case. I mean, come on. I don't think I've gotten all his or her posts yet, but I've posted...um...two direct responses? I think. One was really short, saying "no, you're wrong." One was a tiny li'l flame. Where I *really* took out my anger was in my response to River-- though strangely, it came out not as an attack, but as a way for me to actually consider how I felt about Blue's words. I don't feel guilty for any of it (a nice change!)...what's that phrase, to "suffer fools gladly"? Well, I don't want to. And I *especially* don't want to when they poke their intrusive noses into my newsgroup over and over, stirring up ridiculous trouble.

Now, I guess there's the question of whether people might get distressed over constant flaming of poor, intellect-challenged Blue ("cain't kick no three-legged dawg", as my grandpa might've said), but I think it's important to note that there were a *lot* of rational, cool-headed responses to those posts-- including some of your own.

I know, I know, the guilt's a built-in feature of depression (Blue Moon, that'd be criterion A.7 of a major depressive episode), so maybe telling you this won't help a lot. In which case, as I have been nominated and seconded for Pope, I hereby grant you a plenary indulgence for the literally hundreds of people you help here, every time you open your mouth. Go forth and sin no more--and remember, there's no "thou shalt not speak thy mind".

--charles padgett **Baywatch Barbie and her pet dolphin do not move by themselves**

In general, Usenetters are able, through relatively mild means of social control like those cited above, to minimize the incidence and effect of deviant behaviors. Like Undernetters, the majority of Usenetters are norm-abiding citizens who voluntarily conform to the community-wide rules of the network. They become regulars on a select few newsgroups that are of interest to them, and they take pride in upholding the quality and effectiveness of those groups; they agree to agree. As Lea, Martin, O'Shea and Fung [1992] note: ". . .far from being uninhibited and deregulated behavior that is universally observed, flaming is in fact both radically context-dependent and relatively uncommon in CMC. . ." [p. 89]

Given that flaming and flame wars are relatively rare in Usenet, the question that arises is 'under what circumstances do regulars define behavior as deviant enough to warrant a flame (and what degree of flame) in response?' Usenetters have at their disposal the kill file. If they wanted to, they could place all deviants in that file and avoid their posts in the future. Or they could simply choose to skip over the threads known to contain posts from deviants. That is, when the screen of threads appears, the title, poster name and email address can give the regular clear clues that the posts contained within are undesirable (at least in the instance of the repeat offender such as Farmer Fischer in the rec.pets.cats newsgroup). Unlike interaction in the Undernet, in which an inappropriate statement or action may simply appear one one's public channel screen, interaction in Usenet is selective in this way. Given these circumstances, reading and responding to deviant posts, either individually or collectively, serves some purpose for the individuals involved. The purpose that is served through such activity is to continually re-establish the boundaries and mission of the newsgroup and the identities of the upstanding regulars.

Consider the following example, in which Farmer Fischer has returned to rec.pets.cats and posted an article with several inappropriate comments, the most important of which are 'women have no business in the Internet' and 'Then I take it your titties no longer point toward heaven'. Note that, even though Fischer attempts to call upon a community-wide norm that one should not quote a very long post only to attach a minimum amount of new information at the end, his attempt to do so is rejected by the group regulars. The regulars of rec.pets.cats will not have a known newsgroup enemy quoting the rules of netiquette to them. They 'gang up' on Fischer to the extent possible in an asynchronous CMC system, and ridicule not only his statements, but his incorrect use of grammar and vocabulary. When one's identity is one's posts, the most effective means of attacking that identity is to attack the very competence of the article itself:

Response #1:

dacktyl@winternet.com (Farmer Fischer) wrote:
>Patricia Morgan (morgank@ix.netcom.com) wrote:
>: The question is not do they think, but do they suffer.

>The question is do you think enough to realize you quoted over 100 lines
>from previous articles just to add your profound one-liner. Or are you just
>another bimbo in the game of life? Women have no business in the Internet.

You're correct, women have no business IN the Internet - all those electrons and such flying around would surely muss my hair! I'm sure glad I'm only participating ON it, and sure wish neanderthals like you would get OFF of it (or at least try using the head positioned ABOVE your waist before posting)!

There are a number of reasons a response will show up when its progenitor does not. This makes it difficult to understand a response whose original post has been snipped and I, for one, am often glad the original post was included so I know what's going on..

Lori, Bandit's House Guest

Response #2:

What's this about " Women have no business in (sic) the Internet." ? I was on ARPANET, one of the ancestors of the Internet while I was working at NASA Ames Research Center in California's Silicon Valley. I was on ARPANET, doing business, while assisting scientists in use of the Cray X-MP supercomputer. The Net has certainly changed!
Mary Fowler

Response #3:

Farmer Fischer (dacktyl@winternet.com) wrote:
: Cathi Kietzman (kietzman@caa.mrs.umn.edu) wrote:
: : In article <47g65f$991@blackice.winternet.com>, dacktyl@winternet.com (Farmer Fischer) writes:

: : >The question is do you think enough to realize you quoted over 100 lines
: : >from previous articles just to add your profound one-liner. Or are you just
: : >another bimbo in the game of life? Women have no business in the
: : >Internet.

: : all women in general? or would that be just the bimbo variety?

: Bimbos and split-tails that spell their names with an "i" instead of a "y" Get it, : sugarhips?

Do you guys realize that dacktyl is basically controlling the thread here? I mean shit, this WAS a post about cats being given away, right? not whether or not dacktyl is sane enough to be on the net or not... god.
--
Life: David Bryan | e-mail: videoman@winternet.com | IRC: VideoMan
"Do we really need a goal in life...." "ich bin ein auslander"
"god said let there be light and there was, but then people made
 sunglasses....."

Response #4:

dacktyl@winternet.com (Farmer Fischer) wrote:

>Then I take it your titties no longer point toward heaven!

To all women who are offended by this: The reason a **** posts such comments is because it is the only way he/it can get a woman to interrelate to him/it. He/it is almost certainly not attractive in any way, shape or form to any human female, and can probably only speak to a woman through a medium such as internet. As he/it is probably as interesting as a tray of cat litter, he/it probably can't get a response even within internet unless he/it insults women. You should therefore feel sorry for him/it, just as you would for any other poor, dumb animal, and not respond in an aggressive manner.

Please insert your own suggestions at **** above. 'Man' is obviously inaccurate, and any other word associated with living species is an insult to the species concerned.

Beth
hunter@singnet.com.sg

Response #5:

In article <485br3$htb@blackice.winternet.com>, dacktyl@winternet.com (Farmer Fischer) writes:

>Then I take it your titties no longer point toward heaven!

I've been trying to envision teats (look it up, snoogums) that point toward heaven, by which I gather you mean "the sky". I've never seen any such, unless said person was lying on her back.

In any case, ONE thing is clear. Mary's far too discerning to be upset by the likes of you. Here's your TO DO list for today:

1. Get a clue.
2. Get a life.
3. Don't bother flaming me. I'm far too contemptuous of such people to be angered or upset.

Cici

Flaming and flame wars, then, may be instigated by anyone. When they are started by the outsider, that individual is defined by the regulars as deviant and he or she is dealt with accordingly. However, flame wars may also be started by the regulars themselves, in which case they are seen as an appropriate means of social control under the circumstances. When the threat to the newsgroup is serious enough, ". . .flame wars are the most important means of social constraint on the Usenet system." [Hardy 1993, 8]

Participants on Usenet most definitely see themselves as belonging to a community that is based upon the basic values of efficiency and cooperation. Rules of behavior have been constructed that are intended to allow all members of the community to participate in the interaction, provided that they agree to follow those rules. That these individuals see themselves as members of something larger than themselves is evidenced by the their willingness to punish rule violators in an effort to preserve the integrity of the group.

Just like Undernetters, Usenetters have developed the means by which to define deviant behavior and which types of deviance warrant which types of responses. That is, while there exist some basic rules of behavior, it is understood by the members of Usenet that each definition of deviance is highly situational. Allowances are made for unintended violations of social norms, provided that the individual in question openly acknowledges newbieness. Further, individuals utilize the frameworks of both public newsgroup threads and private email to define someone as deviant or non-deviant. It is through such mutual aligning of thoughts, even though they take place in an asynchronous system, that deviance can be said to be constructed in interaction. Usenetters have developed social control mechanisms ranging from the polite request, to sarcasm and ridicule, to the kill file; they define their newsgroups, through word and deed, as communities and work together to maintain them.

THE FIDONET COMMUNITY

Where deviance and social control are concerned, the Fidonet community can be said to once again bear resemblance to both the Undernet and Usenet communities. That is, there are certain behaviors such as flooding, profanity, advertisements, off-topic posts, and posts which violate the 'spirit' of the conferences, which are defined as deviant and dealt with accordingly. As was the case within the Undernet and Usenet communities, such behaviors are defined largely on a case-by-case basis and are managed with varying forms of social control.

For technical reasons, the community-wide deviant behaviors found in Usenet are not a problem in Fidonet. However, individual members of Fidonet who have the status of sysop or above are capable of engaging in deviant behaviors that, while they do not require a community response as such (from 'regular' users), do require that other members of the Fidonet power structure exert some sort of social control. The types of deviance and social control which might be engaged in by these higher status community members are varied, and the *Policy4* document is a key strategy for defining and dealing with those deviant behaviors.

Policy4 stipulates the two basic rules of Fidonet etiquette--1) Thou shalt not excessively annoy others, and 2) Thou shalt not be too easily annoyed. But beyond this, the *Policy4* [1989] document also explains the processes of defining 'excessively annoying' and resolving problem behaviors:

> The coordinator structure has the responsibility for defining "excessively annoying"... The first step in any dispute between sysops is for the sysops to attempt to communicate directly, at least by netmail, preferably by voice. Any complaint made that has skipped this most basic communication step will be rejected... Filing a formal complaint is not an action which should be taken lightly. Complaints must be accompanied with verifiable evidence, generally copies of messages; a simple word-of-mouth complaint will be dismissed out of hand.
>
> If this fails to resolve the problem, you should complain to your Network Coordinator and the other sysop's Network Coordinator...
>
> If you are having problems with your Network Coordinator and feel that you are not being treated properly, you are entitled to a review of your situation. As with all disputes, the first step is to communicate directly to attempt to resolve the problem. The next step is to contact your Regional Coordinator...If you fail to obtain relief from your Regional Coordinator, you have the right to follow the appeal process described in section 9.5.

A decision made by a coordinator may be appealed to the next level. Appeals must be made within two weeks of the decision which is being appealed. All appeals must follow the chain of command; if levels are skipped the appeal will be dismissed out of hand. . .

Network Coordinator decisions may be appealed to the appropriate Regional Coordinator. Regional Coordinator decisions may be appealed to the appropriate Zone Coordinator. At this point, the Zone Coordinator will make a decision and communicate it to the Regional Coordinators in that zone. This decision may be reversed by a majority vote of the Regional Coordinators. Zone Coordinator decisions may be appealed to the International Coordinator. The International Coordinator will make a decision and communicate it to the Zone Coordinator Council, which may reverse it by majority vote.

A coordinator is required to render a final decision and notify the parties involved within 30 days of the receipt of the complaint or appeal.

. . .Most of Fidonet Policy is interpretive in nature. No one can see what is to come in our rapidly changing environment. Policy itself is only a part of what is used as the ground rules for mediating disputes -- as or more important are the precedents. In order to accommodate this process, case histories may be added to or removed from this document by the International Coordinator, with such a revision subject to reversal by the Zone Coordinator Council. Should Policy be amended in such a way to invalidate a precedent, Policy supersedes said precedent.

Policy4 is an essential strategy for the continued technical and social order of the Fidonet community. The document lays out a specific strategy that individuals must utilize if they expect to exert any social control over behavior they are attempting to define as deviant. It even includes sample cases from the past as examples of what sorts of situations warrant an 'acceptable' definition of deviant behavior.

Of course, when it comes to defining deviant behavior involving higher status members of the Fidonet community, *Policy4* is not the only means of recourse. In addition to working through the chain of command, one can go directly to the regular users as a source of support. In the example below, an individual had experienced various problems with the sysop of a particular

BBS. In an effort to successfully define that sysop as deviant in the minds of other regular users, he posted the following message:

Area: Chatter

Msg#: 21667
From: Anonymous
To: All
Subj: A bad BBS

This documentation is a WARNING to all people who log on to and enjoy BBS's.
DO NOT LOG ON TO THE WORLD WIDE WEDGIE BBS!
And the following story is why:

Topic: The W.W.W. (sysop: Spawn)

Let me paint a picture of why you shouldn't log on here. I'll start from the beginning. I started off at my terminal screen and logged on to the W.W.W. BBS. Once I got to the Main Menu I was logged off because it says that I had 0 minutes today. This can't be right. Yesterday I was a SysOp. It must be some problem in the files... I'll call him up. He is my friend and all.
Hi *Don (not the sysops real name)? It's Ryan. I thought I was a SysOp on the board... what happened? Over the phone came Don's voice "I turned you into a new user." "Why?" I asked. "Well what have YOU ever done for the BBS?" "I came up with the name. And, um, I also let you use my account on the IMPosium(which WAS a paying BBS) to get files." "I don't use the files anymore." He said. Liar. I thought. He still had some of the files from the imposium last time i checked. "AND I'm going to change the name of the BBS." Don said. I won't continue with the conversation, for, when I backed him up into a corner in the conversation, he hung up. Later he said I could be SysOp again if I gave him my account password on another BBS so he could download files. "fine." was my reply. He then proceeded to do exactly what he did to me before. That was strike two. Strike three was when he got me kicked off another bbs, which I was SysOp of. It's a long story on how... so I won't explain it. But here is a list of reasons WHY you shouldn't call his BBS:
1. One day he told me that he would read everyone's E-Mail. While this is not a crime he said he would just read them all for "Fun".
2. The name of the BBS should be called the Back-Stab BBS.

3. Some of the files have viruses in them. Just so he has the sheer pleasure of ruining your system.
4. The message bases are stupid.
5. The SysOp brags a lot. When he really has nothing to brag about for any good reason.
6. His games mess up very easily. You can get a lot of permanent line noise. Meaning it will and can screw up your phone line if you access any of the games.
7. He never complements on other BBS's. He'll give them a bad word. Unless the SysOp belongs to the BBS or if he has to lie for some reason.
8. Oh, yeah, the SysOp lies. As stated above.
9. Ever been black-mailed? You will be if you log on here.
10. The most important thing is that he will try to do anything to make you mad. Just for attention.

I hope this gets his attention.
DO NOT CALL HIS BOARD!
DO NOT CALL THE WORLD WIDE WEDGIE OR ANY OTHER BBS THAT HAS THIS
PHONE NUMBER: (617)846-7096
CALL AT YOUR OWN RISK!
PLEASE UPLOAD THIS FILE TO AS MANY BBS'S AS YOU CAN! WARN OTHERS!
___ Blue Wave/QWK v2.20 [NR]
-!- GEcho 1.00
! Origin: Nautica BBS - Home of ScreamNet 617.442.6071 (1:101/195)

Those activities which constitute 'excessively annoying' behavior in the everyday life of Fidonet are generally the same behaviors which are defined as unacceptable by Undernetters and Usenetters--flooding, profanity, advertisements, off-topic posts, and posts which violate the spirit of the groups. Depending on how the conference regulars define the severity of the violation of these norms, the 'appropriate' punishment will vary. While Chapter Five dealt mainly with newbie behavior which was defined as 'normal' within the course of socialization, this chapter deals with behavior which has been defined as deviant because it seen, given the particular circumstances, as intentional--as behavior that no 'true' newbie would think of (or be capable of) engaging in. In such instances, means of social control move beyond those classified as polite or friendly into the realm of flames of various forms.

Bearing in mind that the Fidonet community prides itself on providing an inexpensive means for members to get together and discuss whatever topics are of interest to them in a relaxed and friendly atmosphere, it is readily apparent why the behaviors mentioned above are classified as deviant. Unless such behaviors are kept under control, the technical efficiency and social stability of the community are threatened. Consider first the problem of flooding. Flooding in Fidonet consists by and large of the same activity as it did in the Undernet and Usenet. One who floods a conference posts the same message (or a series of messages) repeatedly to the same conference, thereby potentially overloading nodes with excessive traffic and rendering the conference unreadable by the regulars.

Just as flooding has the potential to disrupt the technical stability of a conference, profanity has the potential to disrupt the social stability; it violates the relaxed, friendly and polite types of interaction which are an integral part of the Fidonet value system. Consider the following examples. In the first example, an individual has violated the politeness norm, and is reprimanded by a group regular. Even in the regular's response, a response in which
he could have used harsher language, he instead chooses to use acronyms in place of 'bad' language:[4]

Area: Chatter

Msg#: 21191
From: Roy Witt
To: Barb Dorey
Subj: HELP me find long lost friend

BD> What is WITH these damn Yankees? I mean, what is WITH them?
BD> Hmmm..mebbe it's because the racial rioting season's past 'n they're
BD> bored 'n frustrated, 'n their incestuously impaired red neck hormones are
BD> interferin' with what little synaptical activity they have t' begin with.
BD> With no exceptions that I can think of, I think what passes for their
BD> synaptical activity boiled down to soup under the sun last summer. and
BD> PM lives in Hamilton, the cess pool of the nation, so I guess he can be
BD> forgiven for being a freaking traitor who should just jump in the bay and
BD> start swimming south NOW.

Take it to FLAME, nobody here gives a damn about your opinion. Nor do we appreciate your flaming the regulars of the echo....BTW, GTF out of here b4 it's too late to leave with your tail between your legs....If you stay, after the dwarves get done with you, you'll be moose fodder..

* OLX 2.2 TD * Universe's most hazardous job - Klingon Gynecologist.

-!- GEcho/32 1.20/Pro
! Origin: Pacific Rim Information -=- San Diego, CA -=- (1:202/711)

As the conference Rules for Mindless Chatter and Drivel state, all regulars in the group are co-moderators. Thus, each regular takes it upon him or her self to define and control deviant behavior. In other groups, however, there is generally one moderator and, depending upon the size of the group, perhaps a few co-moderators. In those instances, the moderators are ultimately responsible for enforcing a code of behavior. In the following example, co-moderator (Shawn McMahon) responds immediately to inappropriate language in the Conspiracy conference area:

Area: Conspiracy

Msg#: 53
From: Shawn McMahon
To: Dick Roebelt
Subj: Government control

DR> No it doesn't, you asshole. Geez, what a fuckin' lamer you are.

Dick, you will cease this sort of behavior immediately or you will leave my echo immediately. This is your only warning.

Shawn McMahon
Co-moderator

-!-
! Origin: Void Where Prohibited/2 (1:3806/10@fidonet)

In both of these examples, conference regulars or moderators take it upon themselves to defend the values of the group and the other regulars. It is understood that group regulars are to be treated with respect, and that community values should be adhered to. And because Fidonetters feel that 'everyone' should know not to walk about cursing in public interactions, the polite suggestions generally aimed at 'true' newbies are dispensed with.

Another problematic behavior in Fidonet is advertising, the defining of which is highly contextual. If a given advertisement is on-topic and if the group has chosen to allow on-topic ads, the behavior is not deviant. When either of these two conditions is not met, advertising is defined as deviant. For example, the Psychology Conference Rules state that ads which are directed at helping group members improve their psychological and emotional competence are welcomed, provided that they receive the approval of the moderator in advance. On the other hand, the Mindless Chatter and Drivel Conference Rules state that advertising is not acceptable, no matter what the content. These posts are typical responses to advertising in groups where it is not allowed under any circumstances:

Area: Chatter

Msg#: 21281
 From: Prof. Dammit
 To: Jason Oberle
 Subj: DOOM TOURNAMENT
 On (23 Dec 95) Jason Oberle wrote to ** ALL **...

JO>   !DOOM TOURNAMENT AT THE INFRARED ROSE!

You may have doomed yourself...

JO> here are numbers for Infrared Rose:

JO> 536-0000
JO> 536-0001
JO> 536-0002

What? no area code? How are the players in Texas, California, Washington, Ontario, Georgia, Colorado, & parts unknown, going to call-in?

JO> We're gonna have fun now!

Yes, we are... another brainless sysop that doesn't have the sense to check an area before making his bombing run...

JO>   UpTo

You may be "UpTo" your ears in "chastisements" for posting this "off-topic" garbage here... I've left your origin line untouched for any of the attack dwarves that may care to speak with you in netmail about this blatant violation of the echo rules, that you *obviously* did NOT read...

JO> ! Origin: The_Rose (1:138/292)

Do *NOT* apologize. Do *NOT* whine at being subjected to abuse. Do *NOT* argue, complain, flame, or show obvious reluctance to post on-topic. Any of these will only INCREASE the abuse poured forth upon you for posting here in the first place.

This is a "gentle" warning. Other first round messages may be more or less abusive depending on the mood and digestion of the other Co-Moderators here. Thin-skinned messagers do not survive initiations into this weird place you've blundered into. Trust Russ on this, he will tell you this but he sometimes isn't quite as "polite" as I am, and it's likely my message won't reach you in time. You'll probably be an even *more* naughty person and reap the "rewards" of harsh 2nd round sniping. Better look at Roy's or Vernon's version of da Rulz, those are the two posting them most often.

>	The Ol' Know-it-All,
>	Prof. Dammit

... Mindless Chatter: This is NOT a "local" echo

-!- PPoint 1.88
 ! Origin: Mindless Chatter: Enter at your own risk (1:3613/12.75)

And:

Area: Chatter

Msg#: 21183
From: Peter Mcneill
To: Jason Oberle
Subj: DOOM TOURNAMENT

Hello Jason!

Saturday December 23 1995 03:49, Jason Oberle wrote to ** ALL **:

JO>   !DOOM TOURNAMENT AT THE INFRARED ROSE!

No ads allowed, you

Learn about fidonet before you leap in and post. Read *Policy4*. And you are not in the current nodelist so that better be a valid address or I'll netmail yer NC.....

JO> + Origin: The_Rose (1:138/292)

... COMPUTER.COM installed. SEXLIFE.EXE removed from memory.
-!- GoldED 2.50.G0843
! Origin: Magic Dragon Theatre (905)527-0853 Hamilton, On Canada (1:244/116)

These two sample posts invoke multiple strategies of social control common to Fidonet. They use sarcasm and define the deviant as someone who deserves all of the chastisement he gets (the picture above depicts the deviant as a clown, or 'bozo', a common term in both Usenet and Fidonet for individuals who 'are not doing it right'). They invoke *Policy4* and the hierarchical coordinator structure of Fidonet as proof that he is in violation of not only conference norms, but Fidonet norms in general. And they point out that the private communications channel of netmail may also be used to define the situation in their favor should the individual in question continue the deviant behavior.

The final general category of deviant behavior within Fidonet is the posting of 'off topic' messages or those which violate the spirit of the group. As was the case with Undernet categorizations of 'unwanted messages or comments', there are several types of posts which can be classified as off topic. As the example above points out, even advertising can be defined as off topic in a broad sense and therefore, deviant. More commonly, however, posts which are defined as off-topic fall into one of three basic categories: 1) the 'personal' message in a public format; 2) the 'generally stupid' message out of which no sense can be made relative to the group's style; and 3) the genuinely off topic post--those which discuss a subject entirely different from the one which represents the stated purpose of the group.

Consider first the personal message within the public format of the Echo conference. Depending upon the situation, the means by which the regulars deal with this problem will vary. In the example below, because of the information posted, the regulars define the deviant as one who may not have known better and will therefore be given the benefit of the doubt. While the post is considered deviant, the person is not (at least not yet):

Area: Chatter

Msg#: 22029
From: Carolyn Fleming
To: Jamie Nolen
Subj: HI!

JN>AD->Hi there everyone.. I'm new around here so I'm looking to meet
JN>AD->some new people. I'll tell you a bit about me. I'm from Ontario..
JN>AD->About 10minutes from Niagara Falls.. I'm 17 and in grade 12. I
JN>AD->love meeting new people.. so if you're looking for a new friend send
JN>AD->me some mail. I love writing messages!! Hope to hear from
JN>AD->someone soon :)

JN>Hi Amy,
JN>I've been looking for a "penpal" for a while. My REAL name is James,
JN>I'm in 10th grade, gonna be 16 in January, and I'm in West Greenwich,

JN>Rhode Island; I guess it's a couple hundred miles from where you are. I
JN>like animals, alternative and some heavy music, Christmas, and
JN>computers.

Jamie, there are echoes for teenagers that want to write to each other. This REALLY is not the place to do it. Please, don't stay in this echo or a bunch of really sick and twisted adults are going to say some really ugly things to you.

 * OLX 2.2 TD * If this were an actual tagline, it would be funny.
 -!- Platinum Xpress/Wildcat! v1.1
 ! Origin: UnderGround BBS : 502-622-3384 (1:2450/400)

Messages defined as 'generally stupid' within the context of a group are not granted as much leniency as those of a personal nature. This is because all conference regulars understand that there are a good number of conferences which do encourage general chat or personal messages, and that a person might accidentally mistake a group that was not a chat conference for one which was (particularly if they have not lurked for a sufficient amount of time before posting). However, there are no conferences which encourage or excuse posts which are 'generally stupid'. In the case of general stupidity, even the regulars who momentarily engage in it are reprimanded by the others, as the post below demonstrates:

Area: Chatter

Msg#: 21737
From: Pete Hopping
To: David Jonathan Morse
Subj: aardvark abuse and you

-=> Quoting David Jonathan Morse to Lasana Bones <=-
>
DJM> your mama is so fat, she counts her chins with the census bureau!
DJM> Your Daddy is so big, he takes up a whole row at the theatre!

There is only one thing worse than an idiot posting off-topic in Mindless...and that is another person ANSWERING off-topic posts. We want this person to stop this <spack>. Do not encourage it.

... Go straight to the docs. Do not pass GO. Do not collect $200!
___ Blue Wave/QWK v2.12

-!- OverMail v0.82b
! Origin: Wargaming and Living History at Modem Ready! (1:261/1151)

In conference areas devoted to specific topics, social order is threatened by the off-topic post. A fundamental strategy for maintaining social order and group integrity is the naming of the group. Named groups allow for definitions of what types of interactions are acceptable and unacceptable and what posts are topical or off-topic. Without the continued vigilance on the part of moderators and group regulars, conference areas can potentially disintegrate as the 'signal to noise' ratio grows ever smaller. However, simply naming the conference area is not always the clear solution that it should be. That is, while it is self-evident that the National Cooking Echo conference is devoted toward anything dealing with cooking, some group names cannot sustain such clear boundaries of interaction on their own. In these instances, it is often up to the moderator and some core regulars to clarify boundaries.

In the example below, which contains a series of five posts, the moderator of the Conspiracy conference must work to define group boundaries by defining what sorts of activities might legitimately constitute conspiracies. The first post is from Willis Morrow, an individual who is not recognized by any group regulars. After he suggests a possible conspiracy, the moderator and a group regular engage in an analysis of this suggestion over a series of four posts. In the end, the regular agrees with and supports the moderator's definition of the initial post as a 'troll', and thus one which is not worthy of serious discussion by the group:

Area: Conspiracy

Msg#: 12
From: Willis Morrow
To: All

Subj: bosnian leg conspiracy Nyuk Nyuk!

I think there must be a conspiracy among ARTIficiaL LIMB MANUFACTURERs to send our boys to Bosnia and get their legs blown off so the US GOvt has to pick up the expensive inflated tags for their products Nyuk Nyuk!

Who runs the "AL INDUSTRY" in Amerika and where is he now? Nyuk Nyuk!

-!- Renegade v10-05 Exp
! Origin: Greetings from the Data Basement....(303)477-6945 (1:104/614)

And the subsequent discussion:

Area: Conspiracy

Msg#: 19
From: Zorch Frezberg
To: Willis Morrow
Subj: bosnian leg conspiracy Nyuk Nyuk!

In a msg on <Dec 23 12:41>, Willis Morrow of 1:104/614
writes to All:

WM> Who runs the "AL INDUSTRY" in Amerika and where is he now?
WM> Nyuk Nyuk!

Got proof of this?

Otherwise, don't contribute to the noise.

Zorch Frezberg
Moderator, CONSPRCY

-!- msgedsq 2.1
 ! Origin: http://cybergate.com/~net205 +209-251-7529+ Fresno_CA (1:205/1701)

Area: Conspiracy

Msg#: 27
From: Tim Hutzler
To: Zorch Frezberg
Subj: Re: bosnian leg conspira

WM> Who runs the "AL INDUSTRY" in Amerika and where is he now?
WM> Nyuk Nyuk!

ZF> Got proof of this?

ZF> Otherwise, don't contribute to the noise.

Since when are 'conspiracies' ever easy to *prove*?

Sometimes the more absurd ones turn out to be the case.

Perhaps you should ask for 'evidence.'

___ Blue Wave/QWK v2.12

-!- Maximus/2 3.00
 ! Origin: Madman BBS * Chico, California * 916-893-8079 * (1:119/88)

Area: Conspiracy

Msg#: 30
From: Zorch Frezberg
To: Tim Hutzler
Subj: Re: bosnian leg conspira

WM> Who runs the "AL INDUSTRY" in Amerika and where is he now?
Nyuk WM>Nyuk!

ZF> Got proof of this?
ZF> Otherwise, don't contribute to the noise.

TH> Since when are 'conspiracies' ever easy to *prove*?
TH> Sometimes the more absurd ones turn out to be the case.
TH> Perhaps you should ask for 'evidence.'

For a legitimate comment, I would have; however, I felt that this was such an obvious troll as to warrant it.

Zorch Frezberg
Moderator, CONSPRCY

-!- msgedsq 2.1
 ! Origin: http://cybergate.com/~net205 +209-251-7529+ Fresno_CA (1:205/1701)

Area: Conspiracy

Msg#: 33
From: Tim Hutzler
To: Zorch Frezberg
Subj: Re: bosnian leg conspira

WM> Who runs the "AL INDUSTRY" in Amerika and where is he now?
WM> Nyuk Nyuk!

ZF> Got proof of this? Otherwise, don't contribute to the noise.

TH> Since when are 'conspiracies' ever easy to *prove*? Sometimes the
TH> more absurd ones turn out to be the case. Perhaps you should ask for
TH> 'evidence.'

ZF> For a legitimate comment, I would have; however, I felt that this was
ZF> such an obvious troll as to warrant it.

Yes, count me in on the troll, too.

But, we can all tolerate just a little levity at times?

___ Blue Wave/QWK v2.12

-!- Maximus/2 3.00
! Origin: Madman BBS * Chico, California * 916-893-8079 * (1:119/88)

Fidonet community members suffer the same potential threats to social order and personal identity that Undernet and Usenet community members do. Fidonetters see themselves as belonging to a community that is based upon the basic values of friendliness and cooperation. Rules of behavior have been constructed that are intended to allow all members of the community to participate in the interaction, provided that they agree to follow those rules. That these individuals see themselves as members of something larger than themselves is evidenced by the their willingness to punish rule violators in an effort to preserve the integrity of the group.

Not only are conference members willing to punish rule violators, they have developed systems of classifications of deviant behavior and various means for constructing what sorts of punishments are warranted. That is, while there exist some basic rules of behavior, it is understood by the members of Fidonet conferences that each definition of deviance is highly situational. Allowances are made for unintended violations of social norms, provided that the individual in question heeds the warning and conforms to group expectations. Further, individuals utilize the frameworks of both public conference and private netmail to define someone as deviant or non-deviant. It is through such mutual aligning of thoughts that deviance can be said to be constructed in interaction. Fidonetters invoke social control mechanisms ranging from sarcasm and ridicule, to *Policy4* to conference moderators to maintain group identity; they define their conferences, in both word and deed, as communities and they work together to maintain them.

ENDNOTES

[1] Becker constructs a 'typology' of deviance through combining the presence or absence of the action in question with the presence or absence of sanctions. 'Conforming behavior' is that which obeys the rule and which others perceive as obeying the rule. The 'pure deviant' is one who both disobeys the rules and is perceived as doing so. The more interesting cases, in terms of developing an understanding of social order and identity, lie in the 'in between' possibilities: 1) the 'falsely accused', who is seen as having violated the rules, even though he or she has not; and 2) the 'secret deviant', who violates the rules, but no one recognizes it or responds to it as a violation.

² A cancel-bot is a program that sends out cancel messages. Cancel messages are normally sent out by a newsreader in response to a user's request to cancel a message *if* the user was also the original poster of the message. Sites will ignore cancel messages that do not appear to come from the original poster. Cancel-bots work around this restriction by forging header lines that make it look like the original poster sent out the cancel; they'll usually add something like a "Cancelled-By" header line as well, to keep things nominally above-board. Use of a cancel-bot against anything besides 'consensus spam' would probably create a fierce uproar, ending in tears. [news.admin.net-abuse FAQ,]

³ This post, and its responses, were prompted by a series of posts from someone called Blue Moon, who claims to be an expert on the technical and medical aspects of depression. S/he posted articles which simultaneously challenged the knowledge that the regulars in alt.support.depression have about their own conditions and suggested the possibility that the group was going to be dissolved. Both of these points are known to raise the ire of group regulars, who pride themselves on the knowledge they have of their conditions and depend heavily on the group for emotional support and well-being.

⁴ This is yet another example of interaction in which restraint is shown on the part of the 'good guy'; time and again in social situations which critics of CMC suggest contain no means by which to 'constrain' individuals, the individuals in question choose to constrain themselves. BTW stands for 'by the way' and GTF stands for 'get the fuck...'

CHAPTER 7

SOCIAL INEQUALITY IN ON-LINE COMMUNITIES

"The lack of a generally accepted theory of stratification has resulted in a number of competing concepts and definitions. It is apparent that the major concept in the field of stratification is social class. This term, however, encompasses a variety of bold conceptual and operational definitions. . . Definitions of social class have varied from the use of rough income levels. . .to composite measures based upon such characteristics as income, education, occupation or ethnic identity." [Stub 1972, 2]

As was the case with a number of concepts discussed in previous chapters, CMC communities problematize some of the most taken-for-granted assumptions about social inequality. Stratification, commonly defined as 'a system by which a society ranks categories of people in a hierarchy', is problematized in two interrelated ways for text-based realities. First, stratification is usually discussed in terms of social class. In addition to this categorization, the other possibilities are usually race/ethnicity, gender and age. Clearly, these classifications are based upon the presence and assumed relevance of the physical body. Second, such attributes are usually discussed as the 'causes' of stratification, but it is never clearly explained exactly how this occurs.

Social stratification is explained as an institutionalized hierarchy that is permanent, or at least very long-term; it is perpetuated by the institutions of society, legitimated through ideology, and transmitted inter-generationally through the process of socialization. The preeminence of social class as a means for explaining stratification has its roots in the work of Karl Marx. For Marx [1966], classes are clearly set off from one another by distinctive economic conditions, people are conscious of these distinctions and their place in one particular class (or ethnic, gender or age grouping), and these divisions are maintained through limitations placed on interaction between classes.

In CMC, none of these classifications based upon the physical body can be said to be universally relevant; CMC challenges the underlying assumption that, if one can see it, it must be relevant. Because nothing is technically seen in CMC, these categorizations are only relevant in particular situations and for particular individuals, if at all. In CMC, the concept of 'class' is irrelevant, as is the concept of 'race/ethnicity', unless the situation is one that is built around such a definition.[1] Likewise, gender and age can become problematic, but again, it depends on the particular situation. In any event, such body-related matters tend to be inconsequential for the members of these communities.

What is important in these communities in terms of the criteria that have been established for stratification is, fundamentally, knowledge of and ability to conform to the norms and values of those communities. Two elements are central to this knowledge of and ability to conform to community norms and values; length of time of involvement and ability to talk in the accepted formats. In this sense social stratification is the logical extension of deviant behavior.

According to Pfuhl and Henry [1993], "There are two bases on which people may be assigned to the socially constructed status of deviant. One is a person's objectionable behavior, and the other is the display of objectionable traits. The first case refers to being deviant by achievement, and the second to being deviant by ascription." [p. 121] An achieved status is one based on the official meaning of the actor's behavior; it has been deemed inappropriate. In contrast, deviance as an ascribed status rests on the negative meaning assigned by an audience to unavoidable personal traits people appear to possess:

> "These 'offensive' traits contrast with what are promoted as shared values or attributes. Such conditions, even when partially present, often arouse marked feelings of antipathy or fear while the persons who display them come to be regarded as inferior--physically, psychologically, emotionally or morally. Consequently, these physical conditions become 'stigma', signs or attributes that are deeply

discrediting, and those who display them are stigmatized, i.e., they are disqualified from full social acceptance." [p. 122]

CMC collapses ascribed and achieved status into one; ascribed status originates as achieved status because one's performance is one's 'being' in text-based reality. Members of CMC communities have nothing but text with which to assess the attributes of others. This is not to say that entire ranges of attributes are not assigned to individuals based upon purely text-based interaction; they are, and often this is done on the occasion of the 'first impression'. But because the individual is literally nothing until he or she types at the keyboard, there are no 'unavoidable personal traits' in terms of how one usually thinks of them; 'objectionable behavior' comes first, and 'objectionable traits' are assigned accordingly. Ascribed status is the result of behaviors that depend upon the length of time of involvement in the community and an individual's ability to conform to appropriate talk formats (which in and of itself is dependent on length of time of involvement).

The construction of social norms results not only in the general label of deviance, but in a hierarchical ranking of individuals based upon the extent to which they conform or deviate from those norms. Because social stratification involves not just inequality, but power and beliefs, such hierarchical arrangements are seen as 'fair' by those who best 'live up to them'. Those who do not live up to them are classified as 'outsiders', as 'lower'.

Of particular relevance to stratification in CMC communities is the work of Elias and Scotson [1994], The Established and the Outsiders. In general, this work is an analysis of how deviant behavior is constructed and results in a system of stratification based entirely on differences in 'duration of residence' between two groups of people in a particular community:

> One can observe again and again that members of groups which are, in terms of power, stronger than other interdependent groups, think of themselves in human terms as better than the others. How is it done? How do members of a group maintain among themselves the belief that they are not merely more powerful but also better human beings than those of another?. . .A universal regularity of any established-outsider figuration is that the established group attributes to its members superior human characteristics; it excludes all members of the other group. . .the taboo on such contacts is kept alive

by means of social control such as praise-gossip about those who observed it and the threat of blame-gossip against suspected offenders. . . the full armory of group superiority and group contempt was mobilized in the relations between two groups who were different only with regard to the duration of their residence at this place. Here one could see that 'oldness' of association, with all that it implied, was, on its own, able to create the degree of group cohesion, the collective identification, the commonality of norms. . . Exclusion and stigmatization of the outsiders by the established group were thus powerful weapons used by the latter to maintain their identity, to assert their superiority, keeping others firmly in their place. Members of the 'outsider ' group are constructed as having failed at observing the norms and values of the 'established' groups. [pp. xv - xxxiv]

This discussion of inequality relates particularly well to CMC communities, because the key distinction in status that members of such groups make is between the 'regular' and the 'newbie'. These distinctions are made relevant through talk and can and do result in the uneven distribution of resources and privileges among participants as well as uneven exercise of power in social relationships.

THE INTERNET RELAY CHAT COMMUNITY

The stratification system in Undernet basically consists of operators, channel ops, regulars and newbies; one climbs this social ladder over time, as one learns to fit in with IRC culture. Undernet participants explain the process as follows:

[DISP-Cubbi][Cubbi]   In general an Op becomes an Op because they have proven over time to 1) Be around enough to go thru the trouble 2) Seems mature enough, and 3) is willing.
[DISP-Cubbi][Cubbi]   Well, the proving is really just " Do you click with the other people on the channel, not just Ops, but the people. Do you help others, or just sit, Do you care about the people on the channel at all?"

[DISP-Cubbi][Cubbi]   In General we make friends just like real life, except looks, age, and almost sex even are out of the picture. You can tell who is honest, and who is putting up a front.

Or:

[DISP-#cybersex][beowolf] well, think of all the ways in which men have an advantage in real life relationships....
[DISP-#cybersex][beowolf] :....economic power, political power, old boy networks.....
[DISP-#cybersex][beowolf] and think of how much men judge women by their looks (and vice versa)
[DISP-#cybersex][beowolf] Marcia...NONE of that matters here. All that counts is how clever you are with words

Essentially, Undernetters are interested only in one's facility with words-- the ability to talk in the accepted formats of both the private and public channels, and one's commitment to the values and goals of the Undernet. And while the styles of channels may vary from one another, the basic fact remains: one must demonstrate through a commitment of time, effort and attention, one's worthiness of group membership. This is the case regardless of whether one is a channel op, a regular, a newbie or even an Undernet Operator.

Within Undernet society, operators (opers) are individuals who work together within the Undernet User's Committee to maintain the technical integrity of the system; they play no part in individual channel politics. This is despite the fact that they have the technical ability to exert a high degree of control over individual channels by "/killing" users (breaking the network link that connects users to IRC). Instead, opers work to maintain the atmosphere of the Third Place for all IRC users. According to the Rules for Undernet Operators, "An oper should NOT be any different from an ordinary user as far as possible. The almost sole and whole job of an oper should be to keep the net, in particular his or her server, connected to the net. Remember that every time you use your /oper only commands, it makes the position of an oper more desirable, when IRC should not be considered as a power play. It is only a place where people can meet and talk peacefully without one person having some kind of an unfair 'power' over the other."

It is because the potential for abuse of power by Undernet opers is so clear that extensive rules have been set down regarding the conditions under which an oper may use his or her special powers. These rules are designed to minimize the degree of hierarchy within IRC; in keeping with the belief system which revolves around a commitment to a fair system of communications, differentiation in status is kept to a minimum. A system of checks and balances is employed within the peer group of Undernet opers to ensure compliance with those rules, and that system relies heavily on the appropriate use of accounts. [Scott and Lyman 1975]

The /kill command is the ultimate means of social control within the Undernet. In order to utilize it, an oper must give an appropriate justification both to the user being /killed and the other opers. This justification must demonstrate that no lesser means of controlling the offending individual would have been effective, and the victim of the /kill may appeal the action to the User Committee. If it is found that the /kill was not properly justified, then the oper may be penalized as a result.

With the high degree of potential for social control that comes with the position of Undernet oper, there is also a high degree of responsibility. In order to maintain the Third Place atmosphere, Undernet opers are expected to be courteous and helpful to users. For example, an oper is expected to have a broad enough knowledge of IRC that he or she may respond effectively to user questions. If the oper needs to make any technical adjustments to the server he or she works on, then he or she is expected to inform the users of that server in advance.

In terms of stratification, Undernet opers are expected to make every attempt to minimize social distance between themselves and 'regular users'. Because of the potential for a very lop-sided exertion of social power, opers are expected to be helpful and friendly to users, and they are expected to be able to justify any exercise of the /kill command. Users may appeal, through a democratic process, any action that an oper takes against them, and the oper in question is under pressure to justify his or her actions to the Undernet Committee and the Undernet community as a whole. If the oper fails in this respect, the penalties are severe; for repeated offenses, a degradation ceremony, through removal of the 'O line' (oper status), is warranted. [Garfinkel 1956]

Like Undernet opers, channel ops attempt to minimize the social distance between themselves and other channel participants, be they regulars or newbies. However, although this social distance is effectively minimized, one may still observe stratification in the sense that social norms are selectively applied depending on one's status as op, regular or newbie.

How do channel ops go about minimizing social distance between themselves and newbies? Basically, ops assume the responsibility of welcoming newbies to the channel and making them feel at home. They 'approach' obvious newbies with greetings and helpful tips for getting along in the Undernet. Recall the newbie, Dorothy, from Chapter Five:

[DISP-#25plus][SERVER]   dorothy!~dorothy@204.50.107.20 has joined this channel
[DISP-#25plus][kippers]   hi dorothy
[DISP-#25plus][kippers]   kippers welcomes dorothy to the friendly channel ... :)
[DISP-#25plus][dorothy]   thanks for the welcome
[DISP-#25plus][kippers]   you're welcome
[DISP-#25plus][ACTION]   salt waves at dorothy ~
[DISP-#25plus][kippers]   where are you dorothy?
[DISP-#25plus][ACTION]   OrbWeaver says hello also to dorothy
[DISP-#25plus][dorothy]   I'm from Canada
[DISP-#25plus][salt]   in BC maybe
[DISP-#25plus][dorothy]   i'm in Ontario
[DISP-#25plus][salt]   dorothy, are you a little new on this channel?
[DISP-#25plus][dorothy]   how did you guess? Did my newbieness give me away?
[DISP-#25plus][ACTION]   OrbWeaver did not detect overt newbieness.
[DISP-#25plus][dorothy]   what's the channel about?
[DISP-#25plus][salt]   dorothy...you use capital letters....a newbie-sign :)
[DISP-#25plus][dorothy]   thanks, i'll stop doing that
[DISP-#25plus][ACTION]   salt smiles
[DISP-#25plus][ACTION]   Marcia_Br giggles
[DISP-#25plus][ACTION]   OrbWeaver uses capital letters sometimes .. afraid his 6th grade teacher will return from the dead to correct him.
[DISP-#25plus][ACTION]   dorothy laughs
[DISP-#25plus][salt]   dorothy, if you are curious about the channel and the people...look at our home page

[DISP-#25plus][ACTION]   OrbWeaver tries to use lower case .. a unix thing. oops, Unix.
[DISP-#25plus][salt]   hehe orb
[DISP-#25plus][salt]   it's a little guide into the wonderful world of channel 25plus :-D

Here, both salt and kippers are long-time ops in the channel. They greet dorothy immediately when she enters the channel and try to make her feel at home. Further, they direct dorothy to the channel's home page on the World Wide Web, where she can learn more about the channel and its members. OrbWeaver, who is a channel regular but not an op also tries to make dorothy feel at home by joking with her, but it is ultimately the responsibility of the channel ops to provide dorothy with any assistance she may need.

Additionally, channel ops may take an even greater step in removing social distance between themselves and newbies by 'masquerading' as individuals without op privileges in order to avoid intimidating the newbies:

> As you know I started my channel from my experience from my early days on IRC. I am concerned about the new people on IRC. They know that an @ sign means higher status and power than people without it. So if I remove my operator status, I can easier talk to newcomers on the channel, and they ask me more questions about IRC than when I have the @ sign. I have tried both with and without and noticed this difference. Some of my ops have now actually converted and they are also hiding there op status as I am. It was their own choice, I would never force them to make that decision as you know. [private email 7/17/95][2]

The status of 'regular' is somewhere in between op and newbie. Depending upon how people choose to run their individual channels, social distance between ops and regulars may be minimized either by "converting" newbies into regulars or by giving regulars ops status. For example, the individual quoted above prefers to 'convert' people from newbies to regulars by continually striving to make them feel welcome and respected in the channel. Once this is accomplished, he or she minimizes social distance without giving regulars op privileges by 'treating them as one of the group' and down-playing the importance of ops on the channel.

In other instances, however, the status of 'regular' is effectively removed by giving regulars op privileges, thus leaving only the categories of ops and newbies. Recall an example from Chapter Four in which the channel #41plus was started on the Undernet as a splinter from a channel of the same name on EFNet. Under these circumstances, the people who started the channel felt that the channel would be most stable if all trusted regulars had ops. This way, everyone is on equal footing:

[DISP-#heat][Heatgain]   ok...this channel...on the undernet...was started by about 4 or 5 or so of us......because the effnet was getting too crowded....b4 that...it was started on the effnet..
[DISP-#heat][Heatgain]   well....112 regulars that gave me their real names, addresses, phones....there's more than that on the channel, though..
[DISP-#heat][Heatgain]   effnet seems to be full of.....er....rookies......and college kids....very annoying....plus the splits are horrendous..
[DISP-#heat][Heatgain]   marcia...most channels are very immature...that's why people love ours..
[DISP-#heat][Heatgain]   the other ones have...<especially with the college age kids>....kicking and banning going on all the time..
[DISP-#heat][Heatgain]   that's the principle reason we op one another...protection..
[DISP-#heat][Heatgain]   protection...kids....and morons....try to get ops....kick everyone off...or de-op everyone...and sometimes ban them or close the channel.

Of course, if channel members intend to use this strategy, then they must develop a system for deciding which individuals are trusted enough to be worthy of ops. Again, this is where commitment to the Undernet values and ability to talk in the accepted formats comes in:

[DISP-#geezers][miranda]   that is another thing...time is VERY different here
[DISP-#geezers][miranda]   well...a regular may come on anywhere from one to...ten or fifteen times a day
[DISP-#geezers]<Marcia_Br>   ten or fifteen times??
[DISP-#geezers]<Marcia_Br>   that's a lot.

[DISP-#geezers][miranda] put that in the context of normal social interaction...you are spending days and nights with the same people...
[DISP-#geezers]<Marcia_Br> that's true.
[DISP-#geezers][miranda] heat...would you say that is too high?
[DISP-#geezers][Heatgain] 10 to 15....nope..
[DISP-#geezers][Heatgain] i've been on since about 9:30 today,...
[DISP-#geezers]<Marcia_Br> wow. so the regulars are really into it. i mean they're really involved?
[DISP-#geezers][miranda] anyway...a *day* on irc is the same as maybe....three or four days in real life
[DISP-#geezers][miranda] VERY
[DISP-#geezers][miranda] so...if you are an old timer...you may have only been here six or seven months...
[DISP-#geezers][miranda] okay..think of it like this...
[DISP-#geezers][miranda] you get together with a GOOD friend maybe two or three times a week
[DISP-#geezers][miranda] here...you get together with good friends about five or six times a DAY
[DISP-#geezers]<Marcia_Br> so it speeds things up.
[DISP-#geezers][miranda] so...like knowing heat...we feel like old friends, like in real life...it would be years...but here is only months

This point about time in IRC is critical to understanding how community is formed and how social position is achieved. Participants in IRC repeatedly comment upon their ability to get to know people quickly, and in large part this has to do with overt attempts to be sociable and friendly in order to maintain social order. Time is accelerated due to the nature of the interactions, and as a result, channel members can be reasonably certain, in a relatively short period of time, which individuals can be trusted and which ones cannot. Below, Heatgain goes on to describe how he decides whether or not someone deserves ops and the idea that, while there is no 'formal' decision-making process, channel peers are expected to be able to justify giving ops to someone new:

[DISP-#heat]<Marcia_Br> oh. so how do you decide who to op and who not to op?

[DISP-#heat][Heatgain]   who i op...are channel regulars...people i've spoken with and gotten to know..
[DISP-#heat][Heatgain]   if you come on regularly...joke around with us...and we see you a lot...and you don't cause problems...that would make you a regular..
[DISP-#heat][Heatgain]   if someone comes on...that we don't know...or barely know....no ops...
[DISP-#heat][Heatgain]   if someone asks for ops....no ops..
[DISP-#heat]<Marcia_Br>   because you think they have bad motive?
[DISP-#heat][Heatgain]   marcia...totally subjective...if one of us makes an error in judgement...and the new op turns out to be an asshole....one of us will de-op him..
[DISP-#heat][Heatgain]   marcia...yes...if they ask...that's NOT a good sign..
[DISP-#heat][Heatgain]   i stick to my auto-op list....which contains nicks i've known for a while....others make bad judgements sometimes..
[DISP-#heat][Heatgain]   other channels mostly have one or two ops..
[DISP-#heat]<Marcia_Br>   well, like, how did you develop the auto-op list? i mean, there must be ways you can tell or things you look for in someone's personality to know whether to put them on the auto-op list. is that right?
[DISP-#heat][Heatgain]   what i look for in someone to decide whether to op them?
[DISP-#heat][Heatgain]   i look for....no insults to my friends for a couple of weeks...
[DISP-#heat][Heatgain]   i look for....no attempted kicking of banning if someone DOES give them ops..
[DISP-#heat][Heatgain]   i look for....no unwelcome sexual innuendos to my female friends..
[DISP-#heat][Heatgain]   i look for....maturity......
[DISP-#heat][Heatgain]   i look for...acceptance to the rest of the regulars..
[DISP-#heat][Heatgain]   and..i look for...a sense of humor..
[DISP-#heat][Heatgain]   sometimes others will op someone...and i'll de-op them..
[DISP-#heat]<Marcia_Br>   why?
[DISP-#heat][Heatgain]   because......i don't know them...
[DISP-#heat]<Marcia_Br>   oh.

[DISP-#heat]<Marcia_Br> so do all the ops sort of have to agree about who is op'd then?
[DISP-#heat][Heatgain]   marcia...more or less.....if...let's say kojac <a regular>...ops someone i don't know...i might message kojac with "you know that person?"
[DISP-#heat]<Marcia_Br> o.k. got it. so if you're an op, you have to be able to explain yourself to the others?
[DISP-#heat][Heatgain]   marcia...most of the current crop of ops don't pay that much attention...there is a core of us, however...that kinda perform....guard duty..
[DISP-#heat]<Marcia_Br> ah, now that's interesting. who is in the core?
[DISP-#heat][Heatgain]   well....when i say core...in this instance...i mean some who pay attention to that kind of thing....not how high in the hierarchy they are...understand?
[DISP-#heat]<Marcia_Br> yeah:)

This issue of who deserves op privileges is a function not only of time in the channel, but of showing proper deference to those who are ops in terms of how one speaks to them. Recall from Chapter Six the case of nastyboy. Nastyboy attempted to emulate the actions of a channel regular, and in so doing, stepped across the line of what sort of behavior is acceptable for a newbie status. Intimate gestures, such as offering a rose (@}---'--}-,-----), are reserved for individuals who know each other well enough. Hugging is yet another gesture that is acceptable or unacceptable depending on one's status:

[DISP-#25plus][SERVER]   gwhiz!~gwhiz@gwhiz.ts2.ameritel.net has joined this channel
[DISP-#25plus][X-MODE]   Has changed gwhiz's mode to +o
[DISP-#25plus][mireille]   gwhiz!
[DISP-#25plus]<Marcia_Br>   re gwhiz
[DISP-#25plus][Kenny]   hi g.
[DISP-#25plus][salt]   Gwhiz!!!!!!!!!!!!!!!!!
[DISP-#25plus][salt]   ˆ {*{*{*{*ˆ • gwhiz• ˆ *}*}*}*}ˆ
[DISP-#25plus][gwhiz]   <------ hugs
[DISP-#25plus][ACTION]   OrbWeaver is impressed with salt's expert use of { and *'s
[DISP-#25plus][mireille]   salt is very good at hugging :)

```
[DISP-#25plus][salt]        you need practicing e.i. meeting Gwhiz often ;)
[DISP-#25plus][ACTION]      Marcia_Br wonders if she should practice
hugging gwhiz
[DISP-#25plus][ACTION]      salt knows gwhiz is very huggable
[DISP-#25plus][ACTION]      OrbWeaver has just managed to master the
*hug* after 9 months .. will try a {*hug*} in a few more months maybe...
[DISP-#25plus]<Marcia_Br>   {{{{{{{{{{ORB}}}}}}}}}}
```

In this example, salt and gwhiz are both ops and long-time channel members. Salt gives gwhiz a big hug when she enters the channel, but no one else does; others limit the warm greetings to all-caps and/or the use of several !!!!!!!!s. But, once the topic of hugging has become a joke, Marcia can hug OrbWeaver, because they are on the same 'level'. Neither has been on the channel very long.

Beyond gestures, certain topics of discussion and types of interactions are also reserved for those on the same status level. In the example below, an individual from the channel #41plus violates such rules:

```
[DISP-#heat][Heatgain]      i have an example of someone that no one will
give ops to....
[DISP-#heat][Heatgain]      gladiatr...been on like 2 months or so...certainly
enuff to become a regular..
[DISP-#heat][Heatgain]      but....he talks insultingly to the women.....
[DISP-#heat][Heatgain]      and altho...i THINK that's his way of trying to be
funny...
[DISP-#heat][Heatgain]      it's NOT funny..
[DISP-#heat][Heatgain]      just insulting..
[DISP-#heat][Heatgain]      funny...because...some things he says....the
women would find funny maybe if i said it....but not if he says it..
[DISP-#heat][Heatgain]      plus...the women have told him they don't like
it...yet he continues..
[DISP-#heat][Heatgain]      ie:...i can joke about their breasts....or...er...other
parts....he can't..
[DISP-#heat][Heatgain]      plus...you hafta know WHICH female would
banter with you...and which would be insulted..
```

[DISP-#heat][Heatgain]　　you see...what i've learned here....there are no body language....no inflections....no anything but the written word....and...to communicate well...
[DISP-#heat][Heatgain]　　you hafta be VERY precise...
[DISP-#heat][Heatgain]　　because the written word can be interpreted in a myriad of ways..
[DISP-#heat][Heatgain]　　for instance....if i made a joke or said something about your breasts...or said...pussy...or something about your sexual habits...if you respond with silence...or tell me to stop it...i should..
[DISP-#heat][Heatgain]　　but...i wouldn't...unless i knew you better...
[DISP-#heat][Heatgain]　　..if the female starts it...then it's ok..
[DISP-#heat][Heatgain]　　...the same in RL...

So, while there are social norms that all are expected to follow, depending upon the situation and the individuals involved, the rules are selectively invoked. It is in this way that the Undernet community is stratified. A final example serves to clarify this point. Though there is a general rule against profanity in the public channels, this rule is applied selectively, depending on one's status:

[DISP-#geezers][miranda]　there is no blanket code of behavior...well. right now we are in a discussion about the word fuck
[DISP-#geezers]<Marcia_Br>　oh. whether it's acceptable or not
[DISP-#geezers][miranda]　depending on WHO uses it...usually they will be kicked
[DISP-#geezers]<Marcia_Br>　but some people can get away with it?
[DISP-#geezers][miranda]　but, there is a core group who HAS the freedom to use it
[DISP-#geezers][miranda]　yes
[DISP-#geezers]<Marcia_Br>　so, who has the privilege and why?
[DISP-#geezers][miranda]　the privilege is really for the old group..

So, the Undernet community's system of stratification is based largely upon length of time of involvement. Newbies are expected to talk in ways that defer to regulars and ops. They are expected to return to a given channel on a regular basis and to be friendly in order to be accepted and become regulars.

Once they are established as regulars, they have, in the minds of others, committed themselves to the obligation of maintaining the atmosphere of the channel and are expected to behave accordingly. Additional time is required before the regulars will become channel ops, and in some cases, this may never occur.

If one is willing, as a newbie, to accept the help and guidance of regulars and ops, there are no ascribed statuses that limit one's ability to become an op in the channel of one's choice. Even individuals who are initially inept at the communication formats of the Undernet can learn through observation of the more experienced members. What is required is that the individual acquire an understanding of the particulars of the atmosphere and people of a given channel, and that he or she be able to contribute to that atmosphere in a way that is accepted. Of course, if one fails to do this, then one will achieve 'ascribed' status characteristics that Undernetters construct to describe socially unacceptable individuals--the jerks and the immature--those who are defined, quite simply, as incapable of becoming productive members of the Third Place.

THE USENET COMMUNITY

Any discussion of the system of stratification in Usenet must necessarily encompass the aspects of both community-wide (sysop) and within-newsgroup stratification. In both cases, the traditional categories upon which individuals are ranked are rendered irrelevant, and new criteria for the development of social hierarchies are developed. The criteria for social acceptance are a demonstrated commitment to the values and goals of Usenet and an ability to interact in the accepted formats.

Regarding community-wide stratification, it is interesting to note the similarities between Usenet and the Undernet. Where the Undernet has its opers, who are responsible for keeping servers up and running and connected to one another, Usenet has sysops, or sysadmins, who perform the equivalent technical functions for newsgroups. However, Usenet sysops have additional responsibilities, because in Usenet, the fact that all newsgroups are technically interconnected means that they are also socially interconnected. Thus, sysops have the added responsibility of maintaining the social etiquette of the network in cases where violations of that etiquette are a threat to multiple groups.

Because there is no central authority within Usenet that decides how the network will be run, it is up to the literally thousands of sysops at Usenet sites around the world to work together to the best of their ability to maintain the smooth functioning of the community. Each sysop acts as the 'gatekeeper' at his or her site, determining which newsgroups will be propagated through the site. In the community of Usenet then, the position of sysop is potentially a very powerful one, and one in which the potential for abuse of that power is very great.

But just as there were social constraints on the Undernet opers, so there are on Usenet sysops. First, it takes an enormous amount of time and energy to become a Usenet site. According to the How to Become a Usenet Site FAQ [1993], there are five main technical steps to become a site: make the decision, find a site to feed you news and/or mail, get the software, do what it says, and register your site on the network.

The time and money that are involved in becoming a Usenet site are not insignificant. But that is just the beginning of the process. Once one establishes a site and becomes the sysop, there are a number of on-going community responsibilities that one is expected to live up to by other sysops and by the Usenetters who receive their newsfeed from one's site. First, one must have an intimate familiarity with the rules of netiquette contained in the many FAQs on the subject. Beyond that, one is expected to read and participate in the news.* hierarchy, which is dedicated to the discussion of community-wide issues; one must be able to 'play well with others' when it comes to dealing with other sysops. Finally, one must be prepared to deal with ongoing complaints from users regarding spam and other net-abuse. While the sysop is not expected to be the first line of defense against deviant behavior in newsgroups, he or she is expected to be prepared to 'discipline' users when they commit net-abuse.

Sysops who do not adequately respond to problems emanating from their sites, or who do not cooperate in spirit with other sysops on issues of how the community should be 'run', are dealt with accordingly. In some instances, the site will be boycotted by other sysops, meaning that they will refuse to send it the newsfeeds it needs to give to its users. In other instances, these 'renegade' sysops are embarrassed into conforming with the wishes of the group as a whole. In the example below, from the news.admin.misc newsgroup, there is a discussion about whether or not to create a rec.drugs.* hierarchy within the rec.* hierarchy. Here, Mr. Grubor is in the minority. He has posted various articles claiming that Usenet will be responsible for getting children hooked on drugs if they create newsgroups which openly discuss them. He continues to raise the issue, until eventually Bruce Baugh conducts a private email poll of the opinions of others in the group and posts them:

In the end, I had a total of 37 respondents. This is the final summary of their e-mail, which has been deleted. . .Copies of this are going out in e-mail to those who requested it.

1. Are you in favor of the rec.drugs.* hierarchy?

   28 Yes
    1 No
    8 Indifferent

2. Have Mr. Grubor's posts changed your mind?

   14 Yes, made me more favorably inclined
    5 Yes, made me interested in voting against him
   18 No

3. Have they changed your mind on any other issues?

   3 Yes, I now think we really need to get rid of him
   2 Yes, I am more interested in the group creation process
   2 Yes, I am now even more firmly convinced that this part of the world desperately needs a good mental health system
   2 Yes, I am now inclined to favor anything he opposes
   2 Yes, I am now using killfiles and filters
   1 Yes, I now appreciate how cool they look scrolled fast
   1 Yes, I now dislike Americans more
   1 Yes, I now pity all his neighbors
   1 Yes, I now respect good sysadmins more
   1 Yes, I now think Allisat, Boursy, Holden, and Hayes aren't so bad
   1 Yes, I now think better of retroactive abortion
   1 Yes, I now think free Thorazine is an even better idea

1   Yes, I now think better of the other lawyers, who disbarred him
1   Yes, I now wish for a moderated news.groups
1   Yes, I now wish for the Dave Hayes psychic newsreader more strongly
1   Yes, I am now more firmly convinced that Usenet is in serious decline
15 No

Thus the role of sysop is not without its price. There are a tremendous number of ongoing responsibilities involved, and one must prove one's ability to work well with other sysops as well as with one's site users if one expects to succeed. And, unlike the days before the Great Renaming, most Usenet sites today are so well-connected that the abilities of sysops to disallow propagation of newsgroups simply because they do not care for them has been greatly reduced.

Within individual newsgroups, there is even less immediately noticeable stratification than within Undernet channels. The main reason for this is that Usenet newsgroups do not have the equivalent of channel ops who have 'visible' status distinctions. Therefore, one must spend more time in individual newsgroups, observing who posts, on what topics, and how frequently, in order to discern status distinctions. Given enough time, one can distinguish the relative newbies from the regulars, and the 'core' regulars from the 'average' regulars through studying this 'cycle of statement and response'. In his 1995 study, Searching for the Leviathan in Usenet, Richard MacKinnon discusses Usenetters as capable of exercising the potential powers of 'strength' and 'eloquence' in the cycle of statement and response in order to define and substantiate status distinctions among members:

> . . .but a Usenet persona can have strength relative to other personae. . . 'strength' in Usenet is one's ability to write a potent or even vehement statement. . .'Form' . . .comes from the impression one makes on others not with one's physique but with one's words. Even a pseudonym can convey form. . .Eloquence is possibly the most important power in Usenet. . .in a world where words are primary to existence and serve as the sole mode of communication and activity, their importance cannot be exaggerated. [p. 122]

Exactly how these status distinctions are accomplished depends upon the type of newsgroup; newsgroups falling toward the debate-oriented end of the continuum require different talk formats than do those that lie toward the center or opposite end of the continuum. In the talk.origins newsgroup, which is heavily oriented toward debate and argument, one is expected to have a thorough knowledge of all of the information contained within the group's many FAQs regarding evolution and creationism. To demonstrate otherwise is to call attention to one's newbieness. In the post below (previously discussed in Chapter Five), Justin Hardin, a relative newbie to the group, does just this. He demonstrates a lack of both 'strength' and 'eloquence' because he is not familiar enough with the facts of the matter to make a more forceful or well thought out argument in response to Skeptic's post. A regular in the group responds with a well-balanced explanation of the group's mission and the types of arguments not likely to be accepted in the group. In this way, he demonstrates his own 'strength' by establishing that he is a regular (he knows the rules of the group). He warns this newbie not to 'wander down the unsubstantiated posting path', which will inevitably result in his being defined as an outsider:

In article <48e3o7$e9e@lynx.unm.edu>, Justin Hardin <jmhardin@unm.edu> wrote:
>In article <482hsd$9e0@ionews.ionet.net>, skeptic@ionet.net says...

>> A few months ago I started the thread "Creationist Folly". In the original
>> posting I asserted that creationist where merely those who where ignorant
>> of biology. I note that not a single creationist even claimed to have a
>> biological education, they merely argued from their lack of biological
>> knowledge. Do any of you creationist have a sound biological education. I
>> would enjoy debating these issues but I am frankly tired of trying to give
>> creationists a rudimentary biological education so that we can debate.

>Well, I'm not a creationist...exactly...but I believe it's possible. I've studied
>biology all the way into my second year of college, have a sound background
>in the fundamentals of evolution, and guess that you are both patronizing and
>condescending to anyone that doesn't believe in your personal views.
>Creation was the big bang. Evolution was everything after

Then you are probably *not* the kind of Creationist that skeptic@ionet.net was referring to, and who regularly get flamed crispy on talk.origins. The minimal creationism that said "God started and sustains the Universe" is a purely philosophical position, and not controversial (on t.o, at least; a.a is a different story ;-). At any rate, science has nothing to say about it, one way or another.

>Why are Creation and Evolution mutually exclusive? Answer: THEY ARE
>NOT! Please, try and respect the views of others...to disagree is one thing, to
>patronize and condescend is another!

What talk.origins mostly deals with is Young-Earthers, Flood Geologists, Theistic Anti-Evolutionists and such who propose wholesale revisions of Natural History, claim evolution could not have happened, and make other such "scientific" claims. Almost always, they have most of their facts wrong, or their logic is hopeless. Different views should usually be respected -- but many of these people show minimal respect themselves, accusing their opponents of stupidity and fraud and condemning them to hell-fire. Arrogance in support of demonstrably wrong positions deserves what it gets.

Among the established regulars of the talk.origins newsgroup however, the rules of the group are applied somewhat differently. Established regulars in this group may repeatedly attack not only one another's intellectual competence on the issue at hand, but the persons themselves. This sort of behavior would never be tolerated between a newbie and a regular; it is the type of interaction in which only those who have established themselves as knowledgeable regulars may participate. They may violate many Usenet rules of netiquette. For example, it is commonly understood that one should not re-post an entire article only to insert refutations on a point-by-point basis. Yet this is exactly what the regulars of talk.origins do. They may claim that another regular has engaged in 'creative editing' of a post in order to make the argument appear less sound; another clear violation of standard Usenet etiquette. And, in general, they may engage in all manner of personal attacks and insults, which are generally unacceptable in Usenet.

While such behavior would not be tolerated from a newbie, the regulars have earned the right to do so. In fact, in groups which are heavily debate-oriented, it is almost expected that one do so. It is not that there are no rules of netiquette for the regulars, but that they are applied selectively and situationally. As pointed out in Chapter Six, the newsgroup as a whole can still decide that such exchanges are getting out of hand, even considering the atmosphere of the group, and those involved in the exchanges in question are pressured to tone it down. The general rule of thumb is that a newbie should abide by Usenet norms until that time when he or she is established as a regular in a particular group. Once this is accomplished, the individual has more freedom to push the envelope of norms, always bearing in mind that the group may choose to invoke them as the situation warrants.

Contrast this means of achieving status with that of a group which falls toward the center of the 'newsgroup type' continuum. In groups such as this, one must also demonstrate an appropriate knowledge of the facts concerning the group's mission. But in addition, one must demonstrate an ability in a second talk format, the format of personal concern. Were one to enter the rec.pets.cats newsgroup and attempt to interact in the fashion of the talk.origins newsgroup, one would immediately be defined as an outsider, as a potential threat to the group. Consider the following series of posts, in which an individual named Richard Leith demonstrates an adequate knowledge of cats, but neglects the 'personal concern' format. He is admonished by the 'moral entrepreneurs' within the group (Rob, Andy and Cici) that he has broken an important group norm:

In article <rleith-0512950228500001@0.0.0.0>, rleith@interramp.com (Richard C. Leith) wrote:

> Ear mites. Flush them out with oil is good advice. I would use mineral oil.
> Massage it roughly into the ears and the ear membranes. Twice a day for
> three or four days. While not an emergency they can cause ear

roughly? how 'bout thoroughly? and be very gentle if you do this!

or does "C" stand for "keffo"? :-)

-- rob

And:

rleith@interramp.com (Richard C. Leith) wrote:
>Actually, I am not sure what a "keffo" is. I did just start reading this group
>and do not know the lingo.
(snip)
>So I will pick my words more carefully. Not as light a group as it would first
>appear....;)

Hi, Richard,

I'm sure rob didn't mean any insult. it's just that we had a _very_ heated discussion in this group after this 'keffo' guy (his login name) proposed some less than acceptable methods treating a cat. At least _most_ of us love our furballs and don't want to hear any possibly harmful hints (without a smiley specifying 'I'm not serious, folks', anyway).

In general, this *is* a light-hearted and sympathetic group, the only one, BTW, I read regularly. Welcome aboard!

Andy

And finally:

In article <ram-0512950837270001@ram.tiac.net>, ram@tiac.net (robert a. moeser) writes:

>>Ear mites. Flush them out with oil is good advice. I would use mineral oil.
>>Massage it roughly into the ears and the ear membranes. Twice a day for
>>three or four days. While not an emergency they can cause ear

>roughly? how 'bout thoroughly? and be very gentle if you do this!

Perhaps "vigorously" would be a better term. Not that the cat will admit there's a difference.

Cici, the Evil Mom Who Gives The Medicine

In the example above, a newcomer has a possible solution to a cat health problem. However, he does not word the post in a manner which conforms to the group style, thereby indicating that he is both an outsider and a potential threat to the group. In fact, the posted response by Rob points out to other group regulars that they may, in fact, have another "keffo" on their hands. This is a reference to an individual who attempted to post a solution to a cat behavior problem and who also demonstrated an inability to conform to the 'personal concern' format of the group. However, in that instance, the individual in question did not acknowledge newbieness in the 'proper' way, nor did she apologize, even though she was informed that she should:

Post #1:
paradox@mail.utexas.edu (Meredith Thompson) wrote:
>In article <49vc7r$85h@rebecca.albany.edu>, "A. Hutchinson" <ah610@cas.albany.edu> says:

>>keffo <keffo@islandnet.com> wrote:
>>>the peroxide solution is 3% (it's for contact lenses)

>>still dangerous

>This is a little off, but.. I've never heard of putting hydrogen peroxide on
>contact lenses. Saline solution and distilled water, yes. Is this an actual,
>accepted use of hydrogen peroxide?
>meredith

Well, I use a product called "AOSEPT" there are others on the market, but yes, they are basically a very dilute peroxide solution- SPECIALLY made for disinfecting contact lenses. You need a special cup and a "aodisc" which has a coating on it that interacts with the solution and disinfects the lenses over a 7-9 hr period. The solution is NEUTRALIZED and becomes inert at this point. This is the only time it is safe to put in the eyes (and one is supposed to RINSE the lenses anyway). So that is why I told "keffo" that it is still dangerous--Point being, the solution is not safe before it goes through the neutralizing process overnight in the cup/disc.

Thus she/he was spraying a caustic solution on the cat!

I have accidentally put a lens in with this solution (unneutralized) and my eye burned for hrs/even after I rinsed it out with saline. So I speak from EXPERIENCE when I say it is a nasty thing to spray on anyone/any animal!!!

Amber

Post #2:

In article <49qn29$ndq@sanjuan.islandnet.com>, keffo <keffo@islandnet.com>
wrote:

>my name is kristen with an e. why would i lie about using the contact
>solution as a deterrent? if i wanted to impress people i would have come up
>with something better than that. yes, i used this method. only twice, and
>that's all that was needed. the cat no longer goes behind and chews wires. all
>i did was spray him in the face while he was chewing the wires. the bottle
>can be squeezed and made to squirt a fair distance. i don't understand why
> people are all upset about this. what is trolling? listen..i'm new here and not
>trying to bother people. i felt that my method might help other people.
>instead, idiots have been posting things such as "keffer can't spell" and
>"what a troll"..please.. i don't agree with everything you people say, and if i
>complained every time someone spelled a word wrong, we'd have millions
>of postings from me. if you don't like my method, don't use it. simple as
>that.

>I live in fear of not being misunderstood.-- Oscar Wilde

Methinks you probably weren't misunderstood.

Look Kristen when somebody posts an obviously abusive suggestion, us cat lovers who use this newsgroup to share information and help others will assume it's a troll trying to get our dander up.

If you AREN'T a troll, then your suggested solution to cats chewing on things is downright abusive. It's not a question of not liking your method ergo don't use it -- YOU shouldn't be using it. If you really are spraying peroxide in your cat's eyes, you shouldn't have a cat!

I don't care if you misspell a word, I don't think animals should be entrusted in your care. I do care about your lack of understanding why it's WRONG.

Geeze, I feel like I'm listening to another abusive person in the making.

Post #3:

keffo (keffo@islandnet.com) wrote:
:my name is kristen with an e. why would i lie about using the contact solution
:as a deterrent? if i wanted to impress people i would have come up with
:something better than that. yes, i used this method. only twice, and that's all
:that was needed. the cat no longer goes behind and chews wires. all i did was
:spray him in the face while he was chewing the wires. the bottle can be
:squeezed and made to squirt a fair distance. i don't understand why people
are :all upset about this.

People are upset because you have twice deliberately burnt a cat's eyes with hydrogen peroxide. Not only have you bragged about this cruel act, you are suggesting that others do the same.

Your posting of an act of abuse and encouraging others to do the same looks like someone who is posting something offensive in order to draw responses. I, however, noticed that dogged air of `I-am-right' that lingers like a fetid stench over everything you post. If indeed you are 13, that is no excuse for poor verbal skills or animal abuse.

There are other ways to prevent a cat from chewing wires. If you had bothered to read any books on the care of cats, you would have known this. If you had read further, you would have discovered that you could have permanently blinded your cat by spraying a caustic solution in the face.

Your time would be better spent trying to learn about the proper care of cats rather than posting your offensive drivel. You are an offensively ignorant little

girl, and will remain so as long as you keep your mouth open and your eyes shut.
--rep

Post #4:

I have tried twice to communicate with Kristen [Keffo] [by e-mail]. I have found that she is nothing but a foul mouthed, uneducated, little girl who lacks respect for anything: Adults; Education; Culture; Life; etc. I have received nothing but insulting profanity from this child. I explained that IF she wanted to be accepted by this group that she should issue a **Blanket Apology** to the group -- I was told what to do with my "Blanket Apology".

Those who say that we should "Give Keffo A Chance" are wrong: She needs to GROW UP before we even **CONSIDER** giving her a chance!! I suggest that if you have a KILL FILE to simply add KEFFO to it -- she is not worth listening to, nor worth being insulted by. I sometimes wonder if there might not be a way to get in touch with her parents -- they should know that their daughter is in need of psychological treatment.

If Kristen is reading this the only thing we want to see is an APOLOGY to this group. As I pointed out, because a child uses Adult Profanity, does not make the child an Adult!!! It you can not APOLOGIZE to this group, you are obviously not GROWNUP enough to take part in the discussions -- hey YOU MIGHT LEARN SOMETHING!!!! If you think you have a nifty way of doing something, then be prepared to DISCUSS [notice I said "DISCUSS" not "CUSS"] and DEFEND your IDEA; **ALSO BE PREPARED TO SAY, "SORRY, I WAS WRONG"** !!

Robert Ruskin

Post #5:

It seems I was too generous with Kristen. My apologies to all if that does turn out to be true.

Kristen, if you are reading this, blanket apologies occur all the time when people have inadvertently and unintentionally offended others on the 'net. If you do want to participate here, I suggest you swallow your pride, apologize, and explain.

Gwen

In this cycle of statement and response, group regulars re-assert their status relative to Kristen. They define her behavior as unacceptable, as that which violates group norms. In the process, they too engage in the violation of group norms. But these violations are seen as acceptable. For example, they engage in 'inappropriate language', at times shouting at Kristen (all caps) and even attacking her personally (defining her as not only a troll but an animal abuser as well). So, even in a group in which regulars normally adhere to Usenet-wide norms, they are allowed by others in the group to violate those rules when the newsgroup is threatened.

Finally, consider the alt.support.depression newsgroup, which is oriented toward problem -sharing and -solving. In groups such as this, knowledge of the facts is least important (relative to the other two groups discussed above). What is important is ability to talk in the 'personal concern' format. Provided that one is serious about the topic (a criterion for all newsgroups) and displays the appropriate type of concern for those involved, one can become a respected regular in the group. In the example below, a self-acknowledged newbie is welcomed into the group. She or he has 'lurked' for a while, and thus understands the appropriate style of talk:

Satsuki Aizawa wrote:

>Hi, I have been reading this news group, but never posted before. I have
>been having a hard time with my mood and currently debating whether I
>should get on the medication(prozac, perhaps). I'm am worried about the
>possible side effects. And also, I feel that medication is just a way of
>masking a symptom and not *really* dealing with the core issue. If you have
>any opinion about his, please Email me, thanks

Satsuki - First, welcome to the group. You'll discover that this is a very good bunch of people.

What side effects are you worried about? With the exception of a few side effects, most of them passed within the first couple of weeks.

As far as Prozac masking the real cause? I have to agree. Insulin masks the fact that your pancreas doesn't work right anymore. The kicker is that you can either take insulin or pray for your body to get better.

Medication is not a mask. You have fallen into the archaic belief that "your just not trying hard enough." It takes time but try to understand it's not your fault your mood swings.
--

Pete Lienemann Jr.

    It is the responsibility of the regulars in these kinds of groups to welcome the newbies and make them feel at home. Regulars provide answers to medical questions and offer support to those who are uncertain of themselves. When the interaction involves regulars alone, the degree of support (the number of posts and the amount of reciprocal self-disclosure) increases.
    Such cycles of statement and response are the means by which one establishes oneself as a regular in groups dedicated to problem-solving. One must demonstrate a knowledgeable (in the sense of personal experience) concern for the other and the other's problems. In contrast, the example below is one in which Tom does not conform to the appropriate format:

Dear Friends,

First, I am not a psychiatrist or psychologist. I am also not suffering from depression. I am just a person with a caring ear and soul. I am so concerned by people anymore trying to find happiness and purpose in things that cant possibly bring happiness.

Human beings were never designed to find happiness in anything other than spiritually related things. We were not designed to be satisfied by material or even mental things. Man is much more complex than most people think. Unlike the dogs and other animals who are content to just have a bone or warm lap, man was designed to not only provide for himself but to care about his fellow man and to search out the one who made it all possible.

Thus, if you look for happiness in money, sex, possessions prestige we will only come up empty. If, however we look to the One who created us and follow His plan as He designed it we will be so happy we will literally bubble.

What does His plan entail:
1)The needs of others above our own.
2)Seeing that others needs (especially spiritual ones) are met
3)Families take a very high priority (kids are very important to him)
4)Respect for Him (God)

Many of you will probably slam me for my beliefs but that is all right. I love you anyway. He tells us that people will look scornfully upon us.

Your brother
Tom

Posts such as these are a clear violation of the spirit of the group. Though Tom attempts to display concern for members of the group, he remains suspect for two reasons. First, he claims that he does not suffer from depression. For the regulars, this means that he can never 'really' belong. Second, he employs the strategy of religion to try to make his points. Had he read the FAQ, he would be aware that this is a tenuous strategy at best, and one which is not likely to lead to acceptance by the group.

In short, the key status distinction within newsgroups, regardless of what type of group it is, is the degree to which group and Usenet norms are applied. In all newsgroups, the standard Usenet norms apply to interactions involving newbies. However, after one has acquired the status of regular, through the commitment of time, observance of norms and values, and participation in the cycle of statement and response, one may behave more freely, depending on the situation.

As was true for the Undernet, one begins with an achieved status, and this leads eventually to an ascribed status of some type. Either through open admission or inadvertent display, one achieves that status of newbie in the eyes of newsgroup regulars. From there, one has the option as to what one's future achieved/ascribed status will be. As a newbie, if one accepts the assistance of regulars and apologizes for gaffes, one may achieve the status of regular, and eventually, 'core' regular. If however, one chooses to flout the rules of Usenet and the specific group, one achieves that status of outsider in the minds of the regulars.

THE FIDONET COMMUNITY

Like the Undernet and Usenet communities, the system of stratification within the Fidonet community is multifaceted. There are two scopes of interaction in Fidonet; those involving higher status individuals such as sysops and coordinators, and those involving within-conference moderators, regulars and newbies. Several Fidonet documents outline the rights and responsibilities accorded the positions of sysop and coordinator. In order to fully understand the interworkings of this power structure, a brief review of the relevant roles, and their meaning within Fidonet culture, is in order. According to the Fidonet Primer [Schwartz 1995]:

> The nodes which make up Fidonet are owned by individual hobbyists, schools, businesses, newspapers, governments, and clubs. Since most of them are Bulletin Board Systems first, and Fidonet nodes second, they are an independent lot; they always have the option of leaving Fidonet. . . Fidonet consists of: An International Coordinator; ix Zone Coordinators; A few dozen Regional Coordinators; Scores of Net Coordinators; and a large number of Hub Coordinators.
> The IC is elected by the Zone Coordinators from among themselves; the Zone Coordinators are elected by the Regional Coordinators in their Zone; and all of the other Coordinators are appointed by the level above them, and serve at their pleasure. (Note that the Zone Coordinator appoints the very Regional Coordinators who in turn elect him.) The primary duty of each Coordinator is to edit a portion of the Nodelist; that portion is sent up the chain for consolidation and then a master update is passed back down. Their other duty is to settle disputes; their only power to enforce their

decisions is embodied in their control of a Nodelist segment, and that means that the only effective punishment which can be meted out is excommunication (loss of a Nodelist entry). The Network Coordinators have the additional duty of fielding new node applications. . . the diplomatic skills of a Coordinator can make the difference between a happy Net and a Net in open rebellion.

The tension between a rigid autocracy on the one hand and a "go shove it" attitude on the part of the individual sysops is what keeps Fidonet flexible. [pp. 4-5]

Thus, within the Fidonet community, there exists a certain amount of tension with regard to this formal structure and the role it should play in the management of the community. It is this tension, this fundamental suspicion of hierarchical structure within Fidonet, that makes the relationship between the coordinator structure and the everyday user much less lopsided than it might otherwise be.

With the installation of such a structure, the means by which the Fidonet nodelist was developed and disseminated was also changed. Instead of a few individuals being in control of the nodelist, all network coordinators (NCs) are responsible for their portion of it and are required to make it available to users. With the nodelist readily available to all sysops, it is possible for all nodes to communicate with one another without the approval of anyone within the coordinator structure. Thus, in the end, the fundamental goals of Fidonet can survive.

Furthermore, there are expectations implicit in the position descriptions that take into account the original intent of Fidonet. For example, according to the Big Dummy's Guide to Fidonet [Schuyler 1992], the network coordinator is the "host" of the network. The term "host" was chosen to ". . .get away from authoritarian overtones. His responsibilities are to coordinate Fidonet within his network. . .NCs often do lots of things that are beyond what they have to do. . .Ideally, a network operates in a "collegial" atmosphere with everyone in the network contributing to its smooth operation." [p. 3]

Even the position of Regional Coordinator (RC), in which a good deal of potential power is embedded, simultaneously has enough responsibility attached that RCs are generally kept 'in line' by the sheer weight of the role. Regional Coordinators who do not follow the general spirit of Fidonet can easily disrupt the smooth functioning of the network, and can exacerbate any social problems if they are handled ineptly. As a result, RCs tend to put enough pressure on one another such that any 'bad guys' are not in that position of power for very long. Too much is riding on it. As noted by Schuyler [1992]:

> The original purpose of the Regional Coordinator was to be a catch-all network for those nodes which were not close to a network of their own. That's all. But as Fidonet grew, so did the Nodelist, and so did the PIECES of the Nodelist. Since these were all being sent to the Zone Coordinator for incorporation into the master list, this began to be an overwhelming job for the Zone Coordinator.
>
> So someone noticed that the Regional Coordinators didn't have much to do, so why couldn't they coordinate the Nodelist segments from within their region, then send the larger segments to the Zone Coordinator? That way the ZC would have fewer segments to patch together. This would lessen the load, spread it out amongst the regions, and everything would be better. . .
>
> Why was it a mistake? Because it inadvertently created a hierarchical power structure. . .Regional Coordinators are now responsible for appointing Network Coordinators. They are also responsible for the smooth operations of networks within their region. They are responsible for assigning numbers to new networks being formed, and for ensuring that new nodes belong to the right geographic networks. They also are part of the appeal process when a node has problems with a Network Coordinator. You can see from this description that Regional Coordinators now do far more than the original job of taking care of the odd node not belonging to a network. [p. 4]

Interestingly, members of the 'upper' strata of the Fidonet community--sysops and coordinators at all levels--are in positions of potentially extreme power on the one hand, but are potentially irrelevant on the other. As long as they continue to work together to create a smooth technical and social base for the community, their positions are granted a certain amount of prestige and deference by regular users. But, for the technical reasons already discussed, those regular users always have the option of bypassing the coordinator structure should the need arise. The levels of administration are designed to distribute control of Fidonet to the lowest possible level, while allowing for community stability. Should the individuals involved fail to perform adequately, they become irrelevant to the community and may be ignored.

Within individual Echo conferences, Fidonet renders irrelevant the 'real world' classifications by which people usually achieve systems of social differentiation. With flesh-based means of stereotyping unavailable, Fidonetters rank one another by the same two criteria used in the Undernet and Usenet communities; the ability to talk in accepted Fidonet conference formats, and a demonstrated commitment to community values and goals, both of which are acquired only by devoting a sufficient amount of time and energy to learning about the community.

The positions, responsibilities and strategies of stratification within Echo conference areas are remarkably similar to those in Undernet channels in three specific ways. First, there are three general identities within a conference: 1) the moderator (whose parallel in the Undernet is the channel op); 2) the regular (whose ranking lies somewhere between the moderator and the newbie); and 3) the newbie (whose general status is by now understood). Second, conference moderators serve a dual role of welcoming newbies and responding to their questions, and of enforcing the rules of the group (in dealing with both regulars and newbies). Third, the key stratification strategy within the conference is the selective application of norms (the types of interactions that are defined as acceptable), depending on the individual's recognized status in the group.

The conference moderator is the individual ultimately responsibility for the health and well-being of his or her conference. In order to maintain group solidarity, moderators engage in several types of interaction, depending on the circumstances. They try to achieve a fine balance between being in charge, on one hand, and being 'just one of the guys' on the other. How is this accomplished? In CMC, these individuals accomplish multiple roles entirely through the tone of their messages.

In the sample post below, a co-moderator of the UFO Conference area sends a polite but firm warning to a group regular that his response to a previous message was inappropriate. In this instance, the tone of the message is firm, but more importantly, it is signed 'moderator', rather than with the individual's name. This alteration of signatures is a basic strategy by which a moderator may shift tones, and therefore roles, in order to achieve the desired goal:

Area: Fido: Open UFO Discussion & Science Thought

Msg#: 27
From: Glenda Stocks
To: Joe Morris
Subj: Latest On Mass Landings From Ashtar.

JM> Gimme a break.... What's next? dancing bears?

Joe,

Please don't make fun of people.

Thanks,
Moderator

-!- GEcho 1.02+
! Origin: FREQ SEARCHNT.ZIP 508-586-6977 SearchNet HQ- (1:330/201)

The status of 'regular' is somewhere in between moderator and newbie. Depending upon how people choose to run their conference areas, regulars may be a 'step down' from moderators, or they may be relatively equal. When they are relatively equal (which is a decision and circumstance brought about by the moderator), regulars obviously take on the social roles of the moderator as described above, at least to a certain extent. That is, they welcome newbies and enforce the code of conduct, as the following examples illustrate:

Area: FIDO - ST: The Next Generation

Msg#: 673
From: Bill Nichols
To: Matt Oshields
Subj: Cool it, please.

MO>NOW GOD DAMNIT warp 13 is NOT 9.99999... IT'S
MO>13.000000000000000000000000000000000000000000000000

Matt, it's just that you haven't been here very long. You're yelling about a misunderstanding of things that most folks here talked to death quite a good while ago. Those of us who've been here more than a few weeks know what he means, wrong though it might be.

But please do cut out the yelling. Your links are going to get cut if you don't flex some self-control muscles pretty soon. <vbg> This sort of thing is VERY highly frowned upon in Fido.
-!-
þ QMPro 1.53 þ "If you're drunk, I'm funny."

-!- WILDMAIL!/WC v4.11
! Origin: Louisville Hot House (1:2320/180.0)

And:

Area: National Cooking Echo

Msg#: 75
From: Ron Curtis
 To: Burton Ford
Subj: Re: Coffee for me please!

BF> Just wanted to welcome you to the echo. Ergo, Welcome!

I just can't get over just how friendly you all are. I had seven 'Welcomes' today from all over the U.S. Even your very nice Moderator went out of her way to say Hi! and Welcome. Such warmth!

BF>As we say here, pull up a chair to the table and have a cuppa coffee. Or in BF>your case tea? 8-)>

I take mine with cream and two sugars please. Can't stand Tea Ughhh! Though my wife Lilian, wonderful girl, lives on the stuff.

BF> so much fun to read. I am not a cook, but after reading for several

I am, or was, a Chef for many years and did all my cooking by the old 'seat of the pants' method. Throw it in the pot and if it tasted good then both I and the customer was happy. I can't ever remember using a set of scales for anything. You, my friend, would make a great Chef. People who can't cook experiment.

People who experiment discover new ways to serve food. People who can't cook make great Chefs! A friend of mine, another Chef, decided to experiment and threw the following into a pan of hot olive oil. 3 stoned peaches, 1 chopped onion, 1/2lb of lean strips of steak, pinch of mint, wild mushrooms, chopped apple and a can of beer. He simmered it for 1/2 hour then thickened it with a butter/flour mix. Result: Disaster. Next time he left out the mint and it was wonderful and landed him a 'Gold' at a Chef's convention in Leicester.

BF> weeks I ventured to horn in on someone's note (either Wes Pitts the Old
BF> Goat Roper, or Unka Earl, I think). I was stunned and pleased to get a
BF> reply, and have been made to feel welcome ever since - and I struggle to
BF> stay somewhat on topic.

I got in trouble before I started. Didn't know about the brew ban on this echo and posted some the other day. I've dug a trench and am now waiting for the 'fur to start flying'.

BF> Join right in. The moderator Pat is a luv, and the Co-moderator Joanne is
BF> too. Stay near topic (cooking, recipes, restaurants, utensils, books, etc.)
BF> and they will let you enjoy the echo. But don't try to sell anything or
BF> sweet, easy going, angelic Pat will crucify you! 8-)>

Oh Dear! I think I better move before she comes 'gunning for me':-)

BF> 'Til we read again. Burt   /~~.

Nice to talk to you Burt. A real pleasure to meet with you.

BF> * QMPro 1.53 * From the Old Codger. 8-)>

How old is an 'Old Codger' in the States Burt? You sound pretty youthful to me!

Bye for now

P.S. Burt. Lady Di said to say "Hi!" She said she's had enough of the young uns and is looking for a more mature male. Shall I give her your name ;-)

Ron in Blackpool on the NW Coast of Ye Merrie Olde England.

... Without Time, everything would happen at once
-!- FMail/386 1.02
! Origin: Crock's Corner BBS Tel 01253 291023 (2:250/607)

In both of these examples, group regulars take it upon themselves to show the newbies the ropes. In the absence of comments from the group moderators, they attempt to inform the newbie that they 'are not doing it right' and that they should straighten up, before the moderator decides to become involved. From the newbie's perspective, it is always seen as less troubling if a regular can explain the problem, thus eliminating the need to deal with the moderator in his or her 'official' capacity. Such interactions are more likely to spoil one's identity than are advanced warnings from the group regulars.

Because the status of the regular is 'in between', and because it encompasses such a broad range of experience levels (from the new regular to the core regular), the life of the regular within the Echo conference is never cut and dried. While a regular's status is clearly higher than the newbie and somewhat lower than the moderator, any given regular's status vis a vis other regulars is somewhat up for grabs depending upon the situation. As a consequence, what is defined as acceptable behavior within this status group is often negotiated in interaction, more so than are interactions between regulars and newbies.

In the following example, several regulars of the Consumer Report conference area are negotiating the extent to which the violation of a Fidonet norm regarding appropriate quoting and word wrap length on the part of another regular can be defined as deviant and dealt with accordingly. When newbies engage in this sort of annoying behavior, they are reprimanded. When the individual in question is a group regular, however, the decision is not so easy:

Area: Consumer Report

Msg#: 14
From: Wes Leatherock
To: Chuck Fiedler
Subj: Re: NONEXISTENT CONSUMER

-=> Quoting Chuck Fiedler to Butch Gamache <=-

    -=> Quoting Scott Bergeson to BARB JENSEN <=-

BG> AT> By all
BG> AT> means, jump
BG> AT> over to
BG> AT> Consumer
BG> AT> Advocate!

BG> SB> Not to play moderator here, but could you *PLEASE* increase the
BG> SB> line length for word wrap on your quoting?

BG> Irritating, isn't it? I made a comment about it a while back, but I was
BG> ignored. You probably will be too.

CF> Actually, this matter has been discussed several times. Barb claims that
CF> this result comes from her "old computer" which is, of course, silly.
CF> Hardware doesn't determine word wrap, software does. And you don't
CF> need a 32-bit reader to allow you to set word wrap so a change in her
CF> reader would fix the problem. But she prefers to blame it on hardware. At
CF> least that provides a convenient excuse to spend money on a new
CF> machine.   8-)

Barb has mentioned before that she has a Commodore with a 40-column display. Do you have some software you can send her which will cause the quoted matter to display the way you want it to?

Funny...I'm hardly a newbie but I've never found where to set the line length for quoting in my software.

-!- PCBoard (R) v15.22 (OS/2) 10
 ! Origin: Bare Metal BBS - The SouthWest's OS/2 Source (1:147/76)

    In this example, both Scott Bergeson and Butch Gamache are regulars who are thoroughly annoyed that this word wrap problem has been pointed out before but never corrected. Chuck Fiedler, another regular, basically hedges his bets by agreeing that the problem should be fixed, but responding in a polite, almost humorous tone (note the 8-), which represents a smiley with a 'goofy' look). Finally, Wes Leatherock is a regular who defends Barb in her 'absence', pointing out that he is not a newbie, either, and he has never found the appropriate setting on his computer to correct the word wrap problem. Such a comment from a respected group regular serves to absolve the behavior (and the individual) in question from being defined as deviant.

Regulars must also work to negotiate satisfactory definitions of what constitutes appropriate interactions between themselves and moderators. When all regulars are effectively co-moderators, as in the Mindless Chatter and Drivel conference, this is not an issue. However, when regulars are even a small step down in status from moderators, they must continuously work to define their position within the conference community. In an example from Chapter Six, in which a regular of the UFO conference, Joe Morris, is reprimanding a regular whom he feels is not conforming to the spirit of the group, the co-moderator immediately steps in and asks him politely to cease the critique. Another regular, whose name is unknown, takes Joe's side in the issue, and this meets with a response from the chief moderator, Pappas. Note that Pappas' response is directed toward ALL group members, thus pointing out that such challenges are not considered appropriate from anyone in the group:

Area: Fido: Open UFO Discussion & Science Thought

Msg#: 27
From: Glenda Stocks
 To: Joe Morris
Subj: Latest On Mass Landings From Ashtar.

JM> Gimme a break.... What's next? dancing bears?

Joe,

Please don't make fun of people.

Thanks,
Moderator

-!- GEcho 1.02+
 ! Origin: FREQ SEARCHNT.ZIP 508-586-6977 SearchNet HQ- (1:330/201)

Area: Fido: Open UFO Discussion & Science Thought

Msg#: 103
From: Ron Pappas
 To: ALL
Subj: Latest On Mass Landings From Ashtar.

Hi jp,

> > JM> Gimme a break.... What's next? dancing bears?

> >Please don't make fun of people.
> >
> >Thanks,
> >Moderator
> >
>I don't think he's making fun of anyone. I think he's got a pretty valid point.
>There's plenty of proof as to the existence of UFO's, but some of the stuff
>that gets posted to this list is fantasy based on no facts whatsoever. There
>are people who read this list who'll believe anything they are told. I could
>probably make up a pretty decent story that sounds plausible and someone
>would buy into it.

I don't believe it is necessary to belittle or to "poke fun at people" for voicing their opinions. If you disagree on something - then voice your own opinion on the matter with dignity for yourself as well as others.

Peace,

Pap...
The College Board 864.878.7340 FIDO - 1:3639/60

-> Send "subscribe i_ufo-l " to majordomo@world.std.com
-> Posted by: "Ron Pappas" <rpappas@ix.netcom.com>

---SnetMgr 0.60 [r0001]
 ! Origin: SearchNet HQ BBS (508)586-9404 (1:330/201)

To summarize the stratification system within Fidonet, there are three basic status groups to which one may belong--moderator, regular and newbie. Through the socialization process discussed in Chapter Five, one learns that there are advantages and disadvantages to each position. As a newbie, one is somewhat free to ask newbie questions and to make mistakes, provided that one demonstrates his or her willingness and ability to learn the ropes of Fidonet and play well with others. Of course, there are time limitations placed on the newbie; no one will accept that an individual may make newbie mistakes indefinitely. At some point, one is defined as deviant instead.

Regulars are those who have mastered the technical and social skills required to demonstrate their ability and desire to contribute appropriately to the conference community. While they no longer have the freedom to be annoying, as newbies do, they do have other freedoms. They can post off-topic when they need to and can violate the norms of Fidonet, provided that the situation warrants such actions. (For example, they can yell at a newbie for yelling). Additionally, they can contribute to defining the mission of the group, to varying degrees.

Moderators are those who have been around long enough and are thus trusted enough by other group members to have been selected as true leaders. They have the responsibility of ensuring the social survival of the group, but also have the freedom to interact with other members as they choose (provided they do not take advantage of their position). They may switch back and forth between the moderator role and the regular role and thus have quite a bit of leeway in how they may interact.

So, like the Undernet and Usenet systems of stratification, the Fidonet system is based largely upon length of time of involvement. Newbies are expected to read and follow conference Rules, accept help regarding social and technical skills, and talk in ways that defer to regulars and moderators. If they want to be counted as regulars, they are expected to return to a given conference on a regular basis and contribute in what group regulars define as a positive way. Once they are established as regulars, they have, in the minds of others, committed themselves to the obligation of maintaining the atmosphere of the conference and are expected to behave accordingly. Additional time is required before the regulars can become moderators, and in most cases, they never do. Ultimately, through interaction, each individual actively contributes to the negotiation process which determines his or her status.

ENDNOTES

[1] For example, in IRC, there are channels such as #black and #asian, in which people who categorize themselves as members of such groups congregate to 'hang out'. Additionally, in Usenet, there are newsgroups devoted to discussion of economic and ethnicity issues.

[2] When an individual is in an IRC channel, the far right portion of the computer screen contains a listing of all people currently in that channel. Ops have an @ sign before their nick. So, for example, when kippers is in #25plus, her nick at the right appears as @kippers. This way, one can know who is in the channel, and which of those individuals are ops.

CHAPTER 8

AN INTERACTIONIST VIEW OF ON-LINE COMMUNITIES: CONCLUDING REMARKS

The central questions around which this book has revolved are whether communication mediated through technology, specifically computer technology, constitutes so-called real interaction, and whether the establishment of so-called real community is possible solely through computer-mediated communications. In order to answer these questions, the conceptual framework of symbolic interaction and the methodological strategy of participant-observation were employed. The benchmarks of face-to-face interactions were used to examine the commonly accepted understandings of 'meaningful interaction', 'community', and related concepts and to develop the criteria by which these computer-mediated systems could be evaluated.

That is, 'real' interaction and community were defined as experiences deemed meaningful to the participants, as meaningful as face-to-face interaction. Such systems of interaction would also have to simultaneously provide participants with: 1) the means to formulate and maintain individual, legitimated identities; and 2) social institutions, or problem-solving strategies, designed to meet basic community needs (i.e., establishment of a system of values and norms, socialization of new members, resolution of ambiguity, and personal accountability to the group). Failure to pass such a test would mean that, by definition, one or more of these needs remains unmet.

This project was designed to be as comprehensive as possible in scope because, despite a wide range of both anecdotal information and scholarly evidence supporting the notion that the Matrix does allow for meaningful interaction and may in fact be comprised of many human communities, the criteria outlined above had not been utilized to systematically document whether that is in fact the case, or how such communities, if they in fact exist, are developed by their participants. Further, if the computer-mediated social systems could not be said to function as human communities--if the critiques outlined in Chapter One were correct--the sociological grounds for such claims were unclear; the factors essential to community life that were missing from these systems had not been pinpointed.

The reader will recall that critiques of CMC may readily be subsumed under one of four basic and interrelated arguments, which claim the following: 1) interaction on the Matrix is anonymous and thus unlike 'real' social interaction; 2) no one can be held accountable for the statements he or she makes; 3) because there is no formal government, the Matrix is in a state of anarchy; and 4) because of these factors, the Matrix is not comprised of 'real' communities. From a symbolic interactionist perspective, these are inaccurate claims.

Consider first the claim that interaction on the Matrix is anonymous and thus unlike 'real' social interaction. The underlying premise of such an argument is that face-to-face communication: 1) is the appropriate benchmark by which all other forms of communication are to be evaluated; and 2) is not a mediated form of interaction. Setting aside the normative assessment of face-to-face interaction, the argument that such interaction is not mediated is incorrect. A basic tenet of sociological theory is that humans come to define and understand the world through the medium of language. There is no ultimately 'real' world; the world and the self are 'real' insofar as they are known through language. [E. Becker 1962] Thus, at the most basic level, all interaction is mediated through the use of language. Additionally, one could also make the claim that face-to-face interaction is mediated by one's social roles, and by the scene, or frame, in which one is involved.

Because all communication is mediated, there is always the potential for anonymity to some degree. As Goffman [1959] notes, people are 'tricked', fooled, or mislead by another's presentation of self often enough in face-to-face interactions to warrant complex performances in which team members and audience alike work to clearly establish the identity of the other. To put it simply, were the potential for anonymity in face-to-face interactions non-existent, there would be no logical reason for people to invest time and energy into performances to prove to the other who they 'really' are.

The same is true of computer-mediated communications. Clearly, there is always the potential for anonymity among the participants in the same sense that there is potential in face-to-face interactions. But it is not as if the participants in CMC do not recognize this fact. They recognize it as a very clear and present danger to the social order, and because of this, they go to great lengths to minimize anonymity. Basically, participants in CMC have devised cultural norms, rules and tools that allow for and demand a relatively high degree of self-disclosure in order to maintain social order.

In IRC, for example, people select nicks that they feel reveal important aspects of their personality. They freely ask one another questions of a personal nature, particularly in the private channels, in order to establish a sense of security about who the other 'really' is and what they are like. Additionally, they have developed the technical tools, such as the /whois and /finger commands to reveal additional identity information about one another.

In Usenet and Fidonet, the same is true. People sometimes select nicks they feel are appropriate to their personalities; more often, however, they use true names in order to legitimate their identities for the group. They engage in conversations, both in the public spaces of newsgroups and Echo conference areas and in private email, in which they disclose identity-oriented information. They design signature files and taglines in an attempt to convey attitudes and beliefs that they hold.

In every computer-mediated community analyzed, the participants believe that they have sufficient information to determine who others 'really' are to the extent that they decide is relevant for their interactions with them. It should be noted that this critique of CMC as anonymous often implies that an individual's on-line identity may be different from his or her 'real life' identity. The point however, is that all individuals carry multiple identities that are variably relevant, depending upon the context of interaction.

For example, a given individual may have a work identity, a parent identity and a church member identity. However, when that individual is at work, his or her work identity is dominant. The fact that this person also attends church regularly is irrelevant to the situation at hand. The same is true of a comparison between on-line and off-line identities. When an individual joins their CMC community, the identity which is relevant for the other members of that group is the on-line identity. The fact that they are occasionally fooled, just as are people in face-to-face interactions, only serves to increase the amount of work they do to assure one another of their true identities, thus solidifying the boundaries of the group. Through systems of norms and values which demand self-disclosure, and through systems of socialization in which those who are willing to abide by those values and norms are separated from those who are not, participants in computer-mediated interactions have achieved commitment to and reciprocation of individual identities to the extent required for the maintenance of social order.

The second critique of CMC, that no one can be held accountable for the statements he or she makes, is closely tied to the issue of anonymity; no 'real' recourse is possible when no 'real' identity is known. Beyond that, however, it also rests upon the notion that only when the physical body is relevant to interaction are there adequate means of holding people accountable for statements and actions. This is clearly not the case in the computer-mediated communities discussed in this work. Chapters Five and Six described in detail the socialization and social control mechanisms through which new members (and regulars) are held accountable for their conduct within the communities studied.

In IRC, newbies are not allowed by the regulars to lurk around the edges for very long; they are welcomed into the ongoing conversations and are gradually taught, and expected to abide by, the rules of the community. An inadvertent breach of those rules is universally met with a reprimand of some sort, and most often, the newbie apologizes for the slip up and makes an effort to conform to the group expectation. Further, it is not an uncommon occurrence for the newbie (or even the regular) to publicly reprimand themselves before another group member feels compelled to do so. Those who choose not to abide by the rules are defined as deviants and dealt with accordingly. In order to hold such individuals accountable for their actions, group members may /kick or even /ban them from the channel until such time as they choose to become norm-abiding citizens.

In Usenet and Fidonet, lurkers are an unknown quantity and are thus deemed irrelevant to group life. However, as soon as the lurker posts his or her first message, his or her presence becomes 'real' for the group. Thus, they are expected to have read the FAQs or Conference Rules and to be capable of abiding by them. Evidence to the contrary, through inappropriate posts of various sorts, is grounds for reprimand. Again, if the newbie allows him- or herself to be held accountable--if he or she chooses to play by the rules--then he or she may continue to actively participate in the life of the group. Those who refuse to be held accountable are marginalized by the group, either through the use of the killfile or newsfeed filter, or, ultimately, through having their access to the group cut until such time as they prove themselves worthy of membership.

In all instances, participants either individually or collectively invoke the values, norms and technical tools available to them to define and legitimate which actions and statements are considered acceptable and unacceptable; they utilize both public and private means of communication to suggest, define and debate the meaning of a given situation. Once this has been established--when ambiguity has been removed--they employ culturally defined and socially acceptable means of holding all members accountable for their actions. This process includes input from members as well as 'outsiders', as sophisticated means to appeal the definition of the situation have been developed to make community life as fair as possible. It is evident from these facts that community members take very seriously their involvement in these interactions, and that they expect others to take them seriously as well. They have developed the means, in varying degrees of severity, by which to hold one another accountable, and those methods are effective for communities in which the self, even the 'physical' self, is textual.

The third critique, that because there is no formal government the Matrix is in a state of anarchy, was dealt with explicitly in Chapter Four. As previously discussed, it is common practice to define groups and organizations in relation to one another based upon the degree of 'formality' involved; formality is determined by the degree of face-to-face interaction among members and the extent to which rules of behavior are written down as official policy. The 'group' is either primary or secondary, with primary groups being those characterized by intimate face-to-face association and cooperation, and secondary groups, or organizations, being those with 'formal' (written down) rules and regulations and impersonal relationships.

Again, these criteria are problematic for the understanding of CMC communities for two reasons. First, as discussed extensively in Chapters One and Three, these definitions rely on the idea of 'face-to-face' interaction as the standard by which all other communication is to be judged. CMC problematizes this criterion because it de-centers the notion of place; it cannot be said with any certainty whether any type of CMC interaction is face-to-face or whether it is not. This whole notion relies on physical proximity and, as such, is irrelevant to the understanding of CMC groups.

Second, the assumption that written rules are a determining factor in whether or not a given social arrangement is formal or informal is clearly problematic. In a text-based reality, everything is written down one way or another; text is all there is. The only distinction to be made is whether or not rules are written down in specific, issue-related documents or during the course of ongoing conversation. In either case, the members of these communities behave as if these written rules are the policy of the group in question, and enforce and abide by them accordingly.

In Usenet for example, community-wide and newsgroup-specific FAQs have been written and are intended to be what one most commonly thinks of as so-called official policy. Likewise, Fidonet has *Policy4* and each Echo conference has Conference Rules which document the policies of the group. And, even in IRC, a community in which members choose not to create many such documents, there still exist community-wide policies written down in the course of general conversations.

These are important assumptions to challenge because they are in part, as noted above, the standards by which critics of CMC have categorized these phenomena as anarchy. When dealing with computer-mediated interaction, the utility of such distinctions as formal-informal diminishes rapidly. The point is that the people involved in these interactions: 1) act as if they are face-to-face; and 2) act as if the norms of the community, in whatever form they are written, are the official policy. Quite simply, they agree to agree. It is not possible in these interactions to distinguish the primary group from the secondary; the formal from the informal. The individualized groups, be they IRC channels, Usenet newsgroups or Fidonet conference areas are 'primary-enough' in the sense that they are consistently successful in creating new members and socializing them to the rules of the community. Simultaneously, they are 'secondary-enough' (or formal enough) that the rules they develop, by whatever means, are treated as official by the participants; they are consistently and effectively invoked in the socialization and social control processes.

As a consequence, it is impossible to classify CMC as anarchy on the grounds of 'lack of formality'. At a minimum, if one wishes to argue that CMC is anarchy, one must develop new criteria on which to base this argument; the distinction based upon physical proximity and written rules is rendered irrelevant.

But what other sociological criteria are available? It has been established, through the observations of researchers discussed in Chapter One; through electronically available documentation, and through countless comments of participants in CMC, that there are community-wide value systems, some of which are written down in distinct documents with specific titles, others of which are written out during conversation, out of which participants develop elaborate systems of social norms and the means of enforcing them.

This is plainly evident as, time and again, members employ the relevant tools from their cultural toolbox and those tools are recognized and accepted by norm-abiding citizens and deviants alike. Though people may be punished in some way for breaking the rules, no one challenges the fundamental legitimacy of those rules; though they may attempt to define the situation in such a way as to establish that they did not break the rules, or that they had a good reason for doing so, they do not argue that such rules do not or should not exist. Quite simply, there are no means by which one can account for the existence of a system of widely understood and accepted norms, and for the existence of deviance, which is, of course, only definable in relation to such universally understood rules and norms, and still claim that computer-mediated communications systems are anarchy.

As a consequence, the fourth critique of CMC as non-community is baseless, depending as it does on the first three critiques for its foundation. Computer-mediated interactions are not anonymous, they do not lack accountability, and they do not (and could not) take place within anarchy. Large numbers of people have come together within these systems and have worked hard to establish the technical and social means to interact meaningfully with one another. They have succeeded. Every CMC system analyzed in this work has evolved from a piece of communications software into a complex social system, complete with belief systems, sets of values, cultural norms, and systems of socialization, social control and stratification requisite for any collection of individuals to be defined as a human community.

LIST OF ABBREVIATIONS

BBL	Be Back Later
BBIAF	Be Back In A Few (minutes)
BBS	Bulletin Board System
BRB	Be Right Back
BTW	By The Way
CFV	Call For Votes
CMC	Computer-Mediated Communication
CU	See You
EFNet	Eris Free Network (a division of Internet Relay Chat)
F2F	Face to Face
FAQ	Frequently Asked Questions
FITB	Fill In The Blank
FTP	File Transfer Protocol
FUBAR	Fucked Up Beyond All Recognition/Repair
FYI	For Your Information
HNG	Horny Net Geek
HTTP	HyperText Transfer Protocol
IC	I See
IMHO	In My Humble Opinion
IMO	In My Opinion
IP	Internet Protocol
IRC	Internet Relay Chat
IRL	In Real Life
L8R	Later
LOL	Laugh/ing Out Loud
MOO	MUDs, Object Oriented
MorF	Male or Female
MOTOS	Member of the Opposite Sex
MOTSS	Member of the Same Sex

MUD	MultiUser Dungeon
OIC	Oh, I See!
OP	channel OPerator
RE	Repeat Hi/Regards
RFC	Request For Comments
RFD	Request For Discussion
RL	Real Life
ROFL	Rolling On Floor Laughing
RTFM	Read The Fucking Manual
SYSOP	SYStem OPerator
TCP/IP	Transfer Connect Protocol/Internet Protocol
WAN	Wide Area Network
WWW	World Wide Web

LIST OF EMOTICONS

:-)   The Basic Smiley

:->   Ironic or Devious Smiley

;-)   Winking Smiley (for expressions of humor or sarcasm)

:-(The Basic Sad Smiley

:,-(Crying Smiley

:-O   Surprised Smiley

:-D   Smiley With a Very Big Grin

:-p   Smirking Smiley

:-J   Tongue in Cheek Comment

:-V   Shouting Smiley

(name), {name} or [name] -- hugs (big hugs have multiple layers of brackets)

<word>   used to signal physical action, for example <grin>, or <wink>

@>---,---'---- Cyber Rose

<nick> --   user nickname

GLOSSARY

/ban: A basic command utilized in Internet Relay Chat for the permanent removal of deviant individuals from the channel in question.

Bot (or Channel Bot): Short for robot; a standardized Internet Relay Chat program for granting operator status to appropriate individuals within a channel.

Channel: The 'place' in Internet Relay Chat in which synchronous, real-time communications take place.

Clueless: Often used with the word 'newbie', as in 'clueless newbie' to designate one who is a newcomer to a group and does not understand or adhere to the group rules.

Cross Posting: In Usenet, the process of posting a single message to multiple newsgroups.

Echo conference: The 'place' in Fidonet in which asynchronous communications take place concerning defined topics.

EFNet: Eris-Free Network, a division of Internet Relay Chat.

Fidonet: The asynchronous, store-and-forward communications systems linked together in a hierarchical node system and designated by official nodelist numbers.

Flame/FlameBait/FlameWar: Depending on the context, an inflammatory remark which may be designated as either a form of deviant behavior, or a form of social control. Deliberately attempting to start a flame war is known as flame baiting, and flame wars are exchanges of flames which last for an extended period of time.

Flood: The process of sending a volume of text, intentionally or unintentionally, which renders the communication medium (channel, newsgroup, or Echo conference area) unreadable by other participants.

Internet Relay Chat: The synchronous, real-time communications system consisting of multiple interconnected servers, through which participants gain access to individual channels.

Jerk: Designates one who refuses to follow group rules within the channel, newsgroup or Echo conference area in question; distinguished from the newbie or clueless newbie as one who should know better.

/kick: A basic command in Internet Relay Chat used to temporarily remove a deviant individual from the channel in question.

Killfile: In Usenet, a technical device that allows participants to eliminate posts from undesirable individuals from their news feed.

Lag: A condition on Internet Relay Chat in which an individual is unable to sustain real-time communications with other channel members; usually caused by an overloaded server.

Lurker: An individual who observes the interactions of others, but does not participate directly in any of the ongoing conversations.

Moderator: In Usenet and Fidonet, individuals who serve to monitor interactions in newsgroups and Echo conference areas in order to maintain social order.

National Mail Hour: In Fidonet, the designated hour during which appropriate posts are transferred through the node system each night.

Net-Abuse: In Usenet and Fidonet, the term for any of a range of behaviors which threaten the technical and/or social stability of the network.

Netiquette: Net etiquette; the sum total of all socially acceptable rules, values and norms of behavior to which participants are expected to adhere.

Netsplit: In Internet Relay Chat, a condition in which an individual is split off from an overloaded server and finds him or her self in a 'parallel universe'.

Newbie: One who is a newcomer to a group, also refers to one who may be a member but is acting as if s/he does not know the rules.

Newsgroup: In Usenet, the 'place' in which asynchronous communications take place; consists of a set of interconnected servers which continuously transmit posted messages.

Op: In Internet Relay Chat, those individuals who are in charge of a given channel and who have a set of powers reserved for that position (for example, the /ban and /kick commands).

Regular: One who is an established member of the group in question.

Sig: Signature File; in Usenet, designates specific information about the poster of a given message, and can include email address, phone numbers, place of employment, and quotes or sayings.

Spam: In Usenet, designates a message which was cross posted to more than the allowed number of newsgroups; usually results in a universal cancellation of the message.

Tagline: In Fidonet, designates the specific information about the poster of a given message, similar in content to the Signature File in Usenet.

Troll: In Usenet, a message, an activity or a person who deliberately searches out newbies or flamers in an attempt to chastise them; a means of social control.

Twit: A cross between the newbie and the jerk; seen as one who should understand the rules of the group in question, but violates them anyway, although without the malicious intent of the jerk.

Undernet: A division of Internet Relay Chat; started by a group of individuals who decided to split off from the Eris-Free Network by starting their own server system.

Usenet: The asynchronous, continuously forwarding communications system consisting of newsgroups arranged in an hierarchical namespace system, and connected by a system of public and privately controlled servers.

BIBLIOGRAPHY

Adler, Patricia A. and Peter. *Constructions of Deviance: Social Power, Context and Interaction.* Belmont, CA: Wadsworth Publishing Co., 1994.

alt.culture.usenet faq. electronically available at www.cis.ohio-state.edu/hypertext/faq/usenet/usenet/culture-faq/faq.html 1995 (maintained by Tom Seidenberg)

Aluve, Warren. UnderNet User's Committee--Guidelines. electronically available at http://aslan.pr.mala.bc.ca/~warren/usercom_guidelines.html

Anderson, Benedict. *Imagined Communities: Reflections on the Origin and Spread of Nationalism.* New York: Verso, 1991.

Anderson, J.A. and T. P. Meyer. *Mediated Communication: A Social Action Perspective.* Newbury Park, CA: Sage Publications, 1988.

Aycock, Alan and Norman Buchignani. "The E-Mail Murders: Reflections on 'Dead' Letters." In *Cybersociety: Computer-Mediated Communication and Community*, ed. Steven G. Jones, 184-231. Thousand Oaks, CA: Sage Publications, 1995.

Babbie, Earl. *The Practice of Social Research.* Belmont, CA: Wadsworth Publishing Company, 1992.

Basic Information About MUDs and MUDding. electronically available at http://www.lysator.liu.se/mud/faq/faq1.html 1995.

Baym, Nancy K. "The Emergence of Community in Computer Mediated Communication." In *Cybersociety: Computer-Mediated Communication and Community*, ed. Steven G. Jones, 138-163. Thousand Oaks, CA: Sage Publications, 1995.

Becker, Ernest. *The Birth and Death of Meaning.* New York: Free Press, 1962.

Becker, Howard S. Outsiders: *Studies in the Sociology of Deviance.* New York: Free Press, 1963.

Bell, Colin and Howard Newby, eds. *The Sociology of Community*: A Selection of Readings. London: Frank Cass & Co, Ltd., 1974.

Bendix, Reinhard and Seymour Martin Lipset, eds. *Class, Status and Power: Social Stratification in Comparative Perspective.* New York: Free Press, 1966.

Bendtsen, Bo.The FidoNet Homepage. electronically available at: http://www.gpl.net/terminate/fidonet/

Bierstedt, Robert. *Power and Progress: Essays on Sociological Theory,* chapter 13, "An Analysis of Social Power." New York: McGraw-Hill, 1974.

Bierstedt, Robert. *The Social Order.* New York: McGraw-Hill, 1970.

Blumer, Herbert. *Symbolic Interactionism.* Englewood Cliffs, NJ: Prentice-Hall, 1969.

Brissett, Dennis and C. Edgley, eds. *Life as Theater: A Dramaturgical Sourcebook.* Chicago, IL: Aldine Publishing, 1975.

Bruckman, Amy. *Approaches to Managing Deviant Behavior in Virtual Communities.* electronically distributed panel discussion presented at CHI 94 in Boston, MA, April, 1994.

Bruckman, Amy. Identity Workshops: *Emergent Social and Psychological Phenomena in Text-Based Virtual Reality.* Master's thesis, MIT Media Laboratory, 1992.

Bush, Randy. FidoNet: Technology, Use, Tools, and History. electronically available at *gopher://rain.psg.com:70/00/networks/fidonet/inet92.paper* 1993.

Charon, Joel M. *The Meaning of Sociology.* Englewood Cliffs, NJ: Prentice Hall, 1993.

Chesebro, James W. and Donald G. Bonsall. *Computer-Mediated Communication: Human Relationships in a Computerized World.* Tuscaloosa, AL: University of Alabama Press, 1989.

Cohen, Anthony. *The Symbolic Construction of Community.* New York: Tavistock Publications, 1985.

Cooley, Charles Horton. *Social Organization: A Study of the Larger Mind.* New York: Charles Scribner's Sons, 1913.

Copilevitz, Todd. *"Troubled Gather To Sort Out Their Problems On-Line."* The Arizona Republic, 4 January, 1995, D5.

Davis, F. James and Richard Stivers, eds. *The Collective Definition of Deviance.* New York: The Free Press, 1975.

Davis, Kingsley and Wilbert Moore. "Some Principles of Stratification." In *Class, Status and Power: Social Stratification in Comparative*

Perspective, eds. R. Bendix and S. M. Lipset, 47-52. New York: Free Press, 1966.

December, John, *CMC Study Center Resources.* electronically available home page link for papers relevant to computer mediated communication, ongoing and continuously updated.

Dexter, Lewis A. "Introduction." In *People, Society and Mass Communications.* eds. L.A. Dexter and David M. White, 3-28. Glencoe, CA: Free Press, 1964.

Dunlop, Charles, and Rob Kling, eds. *Computerization and Controversy: Value Conflicts and Social Choices.* San Diego, CA: Academic Press Inc., 1991.

Durkheim, Emile. *The Division of Labor in Society.* New York: MacMillan Press, 1933.

Eddings, Joshua. *How the Internet Works.* Emeryville, CA: Ziff-Davis Press, 1994.

Elias, Norbert. "Forward." In *The Sociology of Community.* eds. Colin Bell and Howard Newby, iv-ix. London: Frank Cass & Co, Ltd., 1974.

Elias, Norbert and John L. Scotson. *The Established and the Outsiders: A Sociological Inquiry into Community Problems.* Thousand Oaks, CA: Sage Publications, 1994.

Ellul, Jacques, "Preconceived Ideas About Mediated Information." In *Taking Sides: Clashing Views on Controversial Issues in Mass Media and Society*, eds. J. Hanson and A. Alexander, 344-54. Guilford, CN: Dushkin Publishing Group, 1991.

Farberman, Harvey, and Erich Goode, eds. *Social Reality.* Englewood Cliffs, NJ: Prentice-Hall, 1973.

Fearing, Franklin. "Human Communication." In *People, Society and Mass Communications.* eds. L.A. Dexter and D.M. White, 37-68. Glencoe, CA: Free Press, 1964.

FidoNet Policy4 Document, version 4.06 May 6, 1989. electronically available at: *gopher://rain.psg.com:70/00/networks/fidonet/Policy4.doc*

French, Robert Mills. *The Community: A Comparative Perspective.* Itasca, IL: F.E. Peacock Publishers, Inc., 1969.

Fulk, Janet, Joseph A. Schmitz, and Deanna Schwarz. "The Dynamics of Context-Behavior Interactions in Computer-Mediated Communications." In *Contexts of Computer-Mediated Communication.* ed. Martin Lea, 7-29. New York: Harvester Wheatsheaf, 1992.

Garfinkel, Harold. "Conditions of Successful Degradation Ceremonies." *American Journal of Sociology* 61 (March 1956): 420-24.

Geertz, Clifford. *Local Knowledge: Further Essays in Interpretive Anthropology.* New York: Basic Books, 1983.

Giddens, Anthony. *The Constitution of Society: Outline of a Theory of Structuration.* Los Angeles: University of California Press, 1984.

Goffman, Erving. *Presentation of Self in Everyday Life.* New York: Doubleday, 1959.

_____. *Behavior in Public Places.* New York: Free Press, 1963.

_____. *Stigma: Notes on the Management of Spoiled Identity.* New York: Simon and Schuster, Inc., 1963.

_____. *Interaction Ritual: Essays on Face-to-Face Behavior.* Garden City, NY: Doubleday & Company, 1967.

Gravino, Patrice. *"Heavy Use May Be Creating Cyberjunkies, Researchers Say."* The Arizona Republic, 2 January 1995, E1.

Great Renaming FAQ. electronically available *at http://media2.jmu.edu/users/leebumgarner/gr.html* 1994.

Halloran, S. Michael. "Comment on the Civility Debate." *Computer Mediated Communication Magazine*, 1 July 1995. electronically available at: http://sunsite.unc.edu/cmc/mag/1995/jul/last.html

Harasim, Linda M., ed. *Global Networks: Computers and International Communication.* Cambridge, MA: MIT Press, 1993.

Hardy, Henry Edward. The Usenet System. electronically available at: *gopher://english.hss.cmu.edu/of-2%3a2355%3ahardy-the%20usenet%20system.*

Hauben, Michael. "Exploring New York City's Online Community: A Snapshot of NYC.GENERAL." *Computer Mediated Communication Magazine*, 1 May 1995. electronically available at: http://sunsite.unc.edu/cmc/mag/ 1995/may/haubel.html.

Heim, Michael. "The Nerd in the Noosphere." *Computer Mediated Communication Magazine*, 1 January 1995. electronically available at: http://sunsite.unc.edu/cmc/mag/1995/jan/heim.html.

Hertzler, Joyce O. *Social Institutions.* New York: McGraw-Hill Book Company, Inc., 1929.

Hewitt, John P. *Self and Society: A Symbolic Interactionist Social Psychology.* Boston: Allyn and Bacon, Inc. 1984.

Hiltz, Starr R. and Murray Turoff. *The Network Nation: Human Communication via Computer.* Cambridge, MA: MIT Press, 1993.
Hints on Writing Style for UseNet. electronically available at: *http://www.smartpages.com/faqs/usenet/writing-style/part1/faq.html.*
Homans, George. *The Human Group.* New York: Harcourt, Brace and Company. 1950.
How to Create a New UseNet Newsgroup. electronically available at: *http://www.cis.ohio-state.edu/hypertext/faq/usenet/creating-newsgroups/part1/faq.html.*
Hummon, David M. *Commonplaces: Community Ideology and Identity in American Culture.* Albany, NY: State University of New York Press, 1990.
IRC for the Newcomer. electronically available at: *http://irc.ucdavis.edu/undernet/underfaq/underfaq.1.html 1995.*
IRC Related Resources on the Internet. electronically available at: *http://urth.ascu.buffalo.edu/irc/WWW/ircdocs.html#hyper.*
Jennings, Tom. *FidoNet History and Operation.* electronically available at: gopher://rain.psg.com:70/00/networks/fidonet/fidonethist 1985.
_____. "BBS Etiquette." *FidoNews*, 9 September 1985, 1-4.
_____. "Editorial: Faceless Community." *FidoNews* 16 June 1986, 2..
_____. "Editorial: The War Years." *FidoNews* 24 August 1992, 1-4.
Jones, Steven G., ed. *CyberSociety: Computer-Mediated Communication and Community.* Thousand Oaks, CA: Sage Publications, Inc., 1995.
Judd, Charles Hubbard. *The Psychology of Social Institutions.* New York: The MacMillan Company, 1931.
Kerr, Elaine B. and Starr Roxanne Hiltz. *Computer-Mediated Communication Systems: Status and Evaluation.* San Francisco, CA: Academic Press, 1982.
Kiesler, Sara, Jane Siegal and Timothy W. McGuire. "Social Psychological Aspects of Computer Mediated Communication." In *Computerization and Controversy: Value Conflicts and Social Choices.* eds. Charles Dunlop and Rob Kling, 330-49. San Diego, CA: Academic Press, Inc., 1991.
Kollock, Peter and Jodi O'Brien. *The Production of Reality: Essays and Readings in Social Psychology.* Thousand Oaks, CA: Pine Forge Press, 1994.

Kollock, Peter and Marc Smith. Managing the Virtual Commons: Cooperation and Conflict in *Computer Communities.* electronically available at: http://www.sscnet.ucla.edu/soc/csoc/vcommons.html

Krol, Ed. *The Whole Internet User's Guide and Catalogue.* Sebastopol, CA: O'Reilly and Associates, 1992.

Lauer, Robert and W. Handle. Social Psychology: *The Theory and Application of Symbolic Interaction.* Englewood Cliffs, NJ: Prentice-Hall, 1983.

Laurel, Brenda. *Computers as Theater.* Menlo Park, CA: Addison-Wesley, 1991.

Lawley, Elizabeth Lane. *The Sociology of Culture in Computer-Mediated Communication.* electronically available at: http://www.well.com/user/hlr/vircom/index.html 1994.

Lea, Martin, Tim O'Shea, et. al. "'Flaming' in Computer-Mediated Communication." In Contexts and Computer-Mediated Communication, ed. Martin Lea, 89-112. New York: Harvester Wheatsheaf, 1992

Lewis, Chris and Jonathan Kamens. *How to Become a USENET Site.,* electronically available at: ftp://rtfm.mit.edu/pub/usenet/news.answers/usenet/site-setup

List of Moderators for UseNet. electronically available at *http://www.cis.ohio-state.edu/hypertext/faq/usenet/moderator-list/part1/faq.html*

"Love on the LIne: Pair Meet on Internet, Wed." *The Arizona Republic.* 3 January 1995, A4.

Lynd, Robert S. and Helen Merrell. *Middletown: A Study in Modern American Culture.* New York: Harcourt, Brace and World, Inc. 1929.

MacDonald, Dwight. "A Theory of Mass Culture." In *Mass Media and Mass Man,* ed. Alan Casty, 12-23. New York: Holt, Rinehart & Winston, Inc., 1968.

Macionis, John J. *Sociology.* Englewood Cliffs, NJ: Prentice Hall, 1993.

MacKinnon, Richard C. "Searching for the Leviathan in Usenet." In Cybersociety: *Computer-Mediated Communication and Community,* ed. Steven G. Jones, 112-37. Thousand Oaks, CA: Sage Publications, 1995.

Manis, Jerome G. and Bernard N. Meltzer. *Symbolic Interaction: A Reader in Social Psychology.* Boston, MA: Allyn and Bacon, Inc., 1972.

Martindale, Don. *Social Life and Cultural Change.* New York: D. Van Nostrand Company, 1962.

Martindale, Don. *Institutions, Organizations and Mass Society*. Boston, MA: Houghton Mifflin Company, 1966.

Matheson, Kimberly. "Women and Computer Technology: Communicating for Herself." In *Contexts of Computer-Mediated Communication*, ed. Martin Lea, 66-88 New York: Harvester-Wheatsheaf, 1992.

McCartney, Scott and Joan Rigdon. "Society's Subcultures Meet by Modem." *The Wall Street Journal*. 8 December 1994, B1.

McLaughlin, Margaret L., Kerry Osborne and Christine B. Smith. "Standards of Conduct on UseNet." In *Cybersociety: Computer-Mediated Communication and Community*, ed. Steven G. Jones, 90-111. Thousand Oaks, CA: Sage Publications, 1995.

McQuail, Denis. *Mass Communication Theory: An Introduction*. Thousand Oaks, CA: Sage Publications, 1994.

Mead, George Herbert. *Mind Self and Society*. Chicago: University of Chicago Press, 1934.

Mendelsohn, Harold. "Sociological Perspectives on the Study of Mass Communications." In *People, Society and Mass Communications*, eds. L.A. Dexter and D.M. White, 29-36. Free Press of Glencoe, 1964.

Mills, C. Wright. "Some Effects of Mass Media." In *Mass Media and Mass Man*, ed. Alan Casty, 32-4. New York: Holt Rinehart and Winston, Inc., 1968.

Minar, David W. and Scott Greer, eds. *The Concept of Community*. Chicago: Aldine Publishing Company, 1969.

Mirashi, Mandar. *The History of the UnderNet*. electronically available at: http://sunsite.unc.edu/pub/academic/communications/irc/undernet/

Monberg, John. "Welcome to the Emerald City! Please Ignore the Man Behind the Curtain." *Computer Mediated Communication Magazine*. 1 November 1994, 5-9. electronically available at: http://sunsite.unc.edu/cmc/mag/ 1994/nov/emerald.html

Moraes, Mark. *Hints on Writing Style for UseNet*. electronically available at: http://www.smartpages.com/faqs/usenet/writing-style/part1/faq.html

Moraes, Mark. *Rules for Posting to UseNet*. electronically available at: http://www.smartpages.com/faqs/usenet/posting-rules/part1/faq.html

Mossberg, Walter S. "Accountability Is Key to Democracy In the On-Line World." *The Wall Street Journal*. 26 January 1995, B1..

Ness, Carol. "Females Find Computer Niche." *The Arizona Republic*. 20 December 1994, E1.

Net.Legends FAQ. electronically available at:
http://www.shadow.net/~proub/ net.legends/index.html#kibo
news.admin.net-abuse FAQ. electronically available at:
http://www-sc.ucssc.indiana.edu/~scotty/acena.html
Nisbet, Robert A. *Community and Power.* New York: Galaxy Books, 1962.
Oikarinen, J. and D. Reed. Network Working Group Request for Comments 1459: *Internet Relay Chat Protocol.* electronically available at: ftp://ds.internic.net/rfc/rfc1459.txt.
Oldenburg, Ray. *The Great Good Place: Cafes, Coffee Shops, Community Centers, Beauty Parlors, General Stores, Bars, Hangouts, and How They Get You Through the Day.* New York: Paragon House, 1991.
Parsons, Talcot. *The Social System.* Glencoe, IL: Free Press, 1951.
Perinbanayagam, R.S. *Signifying Acts: Structure and Meaning in Everyday Life.* Carbondale, IL: Southern Illinois University Press, 1985.
Perrolle, Judith A. "Conversations and Trust in Computer Interfaces." In *Computerization and Controversy: Value Conflicts and Social Choices,* eds. Charles Dunlop and Rob Kling San Diego, 350-63. CA: Academic Press, Inc., 1991.
Pfuhl, Erdwin H. and Stuart Henry. *The Deviance Process.* New York: Aldyne deGruyter, 1993.
Priven, Aaron. "NaughtNet: Another New Network." *FidoNews* 8 February 1988, 6-10.
Quarterman, John S. *Which Network and Why It Matters.* electronically available at: gopher://akasha.tic.com:70/00/matrix/news/v1/which.105.
Rapaport, Matthew. *Computer Mediated Communications: Bulletin Boards, Computer Conferencing, Electronic Mail and Information Retrieval.* New York: John Wiley & Sons, Inc., 1991.
Reid, Elisabeth. *Electropolis: Communications and Community on Internet Relay Chat.* electronically available at: http://www.well.com/user/hlr/ vircom/index.html 1991.
Rheingold, Howard. *The Virtual Community: Homesteading on the Electronic Frontier.* New York: HarperCollins, 1993.
Ritzer, George. *Contemporary Sociological Theory.* New York: McGraw-Hill, Inc., 1992.
Robbins, Marge. *FidoNet History Project.* electronically available at: ftp://ftp.netins.net/showcase/fidonet/.web/.index.html

Rose, Donald. *Minding Your Cyber-Manners on the Internet.* Indianapolis: Alpha Books, 1994.

Rosenberg, Bernard. "Mass Culture Revisited I." In *Mass Culture Revisited,* eds. Rosenberg, B. and D.M. White, 3-12. New York: Van Nostrand Reinhold Company, 1971.

Rules for Establishing New IRC Servers in Europe. electronically available at: *http://www.funet.fi/~irc/ebic-rules.html*

Rules for IRC Networking. electronically available at: *http://www.funet.fi/~irc/rules*

Rules for Posting to UseNet. electronically available at: *http://www.smartpages. com/faqs/usenet/posting-rules/part1/faq.html*

Schuyler, Michael. *The Big Dummy's Guide to Fidonet.* electronically available at: ftp://information/Reference/The%20BIG%20DUMMY%27S%20GUIDE%20TO%20FIDONET/Fidonet.1 1992.

Schwartz, Jerry. *A FidoNet Primer.* electronically available at: http://owls.com/ ~jerrys/fidonet.html 1995.

Scott, Marvin B. and Stanford Lyman. "Accounts." In *Life as Theater: A Dramaturgical Sourcebook*, eds. Dennis Brissett and Charles Edgley, 171-91. Chicago, IL: Aldine Publishing Company, 1975.

Shepherd, Clovis R. *Small Groups: Some Sociological Perspectives.* San Francisco: Chandler Publishing Co., 1964.

Smith, Marc. *"Voices from the WELL: The Logic of the Virtual Commons."* Master's Thesis, Department of Sociology, UCLA, 1992.

Snow, Robert P. *Creating Media Culture.* Beverly Hills. CA: Sage Publications, 1983.

Spears, Russell and Martin Lea. "Social Influence and the Influence of the 'Social' in Computer-Mediated Communication." In *Contexts of Computer-Mediated Communication,* ed. Martin Lea, 30-65. New York: Harvester Wheatsheaf, 1992.

Stacey, Margaret. "The Myth of Community Studies." In The Sociology of Community, eds. Colin Bell and Howard Newby, 13-26. London: Frank Cass & Co., Ltd., 1974.

Stein, Maurice R. *The Eclipse of Community: An Interpretation of American Studies.* New York: Harper & Row Publishers, 1964.

Sterling, Bruce. "A Short History of the Internet." Magazine of *Fantasy and Science Fiction*. February 1993, electronically available at: http://www.well.com/user/hlr/vircom/index.html.
Stone, Allucquere Roseanne. "Will the Real Body Please Stand Up? Boundary Stories about Virtual Cultures." In *Cyberspace: First Steps*, ed. Michael Benedikt, 81-118. Cambridge, MA: MIT Press, 1991.
Stone, Gregory and Harvey Farberman. *Social Psychology Through Symbolic Interaction*. New York: Ginn-Blaisdell, 1970.
Suttles, Gerald D. *The Social Construction of Communities*. Chicago: University of Chicago Press, 1972.
Tamosaitis, Nancy. *Net.Talk*. Emeryville, CA: Ziff-Davis Press, 1994.
Templeton, Brad. *Dear Emily Postnews*. electronically available at: http://www.clari.net/brad/emily.html
Tonnies, Ferdinand. *Community and Society*. New York: Harper & Row Publishers, 1957.
UnderNet IRC FAQ. electronically available at: *http://irc.ucdavis.edu/undernet/underfaq/underfaq.1.html*
United States Board of IRC Coordinators Charter. electronically available at: *http://sunsite.unc.edu/pub/academic/communications/irc/efnet/usbic*
UseNet Group Mentors. electronically available at: *http://www.amdahl.com/ext /uvv/group-mentors.html*
UseNet Newsgroup Creation Companion. electronically available at: *http://www.cis.ohio-state.edu/hypertext/faq/usenet/creating-newsgroups/helper/faq.html*
The UseNet Site Administrator's Guide to Netiquette. electronically available at: *http://scwww.ucs.indiana.edu/faq/usagn/*
UseNet Volunteer Votetakers Information Center. electronically available at: *http://www.amdahl.com/ext/uvv/*
van der Haag, Ernest. "Of Happiness and of Despair We Have No Measure." In *Mass Media and Mass Man*, ed. Alan Casty, 5-11. New York: Holt, Rinehart & Winston, Inc., 1968.
Vidich, Arthur J. and Joseph Bensman. *Small Town in Mass Society: Class, Power and Religion in a Rural Community*. Garden City, NY: Anchor Books, 1958.
Von Rospach, Chuq. *A Primer on How to Work With the Usenet Community*. electronically available at: http://www.cis.ohio-state.edu/hypertext/faq/usenet-primer/part1/faq.html

Weber, Max in Schroeder, Ralph. *Max Weber and the Sociology of Culture.* Newbury Park, CA: Sage Publications, 1992.
What is Usenet. electronically available at: *http://www.smartpages.com/faqs/ usenet/what-is/part1/faq.html 1995.*
What is Usenet, A Second Opinion. electronically available at: *http://www.smartpages.com/faqs/usenet/what-is/part2/faq.html*
White, David Manning. "Mass Culture Revisited II." In *Mass Culture Revisited*, eds. B. Rosenberg and D.M. White, 13-24. New York: Van Nostrand Reinhold, Co., 1971.
Whyte, William Foote. *Street Corner Society.* Chicago: University of Chicago Press, 1955.
Whyte, William H. *City: Rediscovering the Center.* New York: Doubleday, 1988.
Wolff, Michael. *NetChat: Your Guide to the Debates, Parties, and Pick-Up Places on the Electronic Highway.* New York: Random House, 1994.
Woolley, David R. "PLATO: The Emergence of On-LIne Community." *Computer Mediated Communication Magazine.* 1 July 1994, 5-16. electronically available at: http://sunsite.unc.edu/cmc/mag/1994/july/plato.html
Wright, Erik Olin. *The Debate on Classes.* New York: Verso Publications, 1989.
Ziegler, Bart and Jared Sandberg. "On-Line Snits Fomenting Public Storms." *The Wall Street Journal.* 22 December 1994, B1.

INDEX

A

action commands, 24, 27, 28
Advertising, 93
anarchy, 2, 31, 32, 33, 34, 38, 270, 273, 274, 275
ARPANET, 13, 14, 15, 207

B

BBS node system, 87, 88, 90, 101
BBSs, 42, 87, 89, 90, 164
beeping, 173
Big Seven, 78, 79, 80, 86
Bots, 120
boycott, 81, 197

C

Call For Votes, 75, 82, 277
channel names, 63, 78
channel operators, 28, 43, 66
channel ops, 49, 65, 70, 176, 230, 232, 233, 234, 241, 244
channel wars, 49
channels, 6, 10, 14, 27, 28, 29, 30, 43, 49, 50, 57, 58, 60, 61, 63, 67, 71, 82, 89, 91, 105, 110, 112, 114, 116, 120, 126, 130, 132, 141, 144, 170, 172, 177, 178, 231, 234, 235, 237, 244, 259, 268, 274, 282
chat-oriented groups, 156, 160
church member, 271
CMC, 7, 8, 9, 10, 11, 22, 30, 40, 46, 47, 48, 82, 89, 169, 205, 206, 226, 227, 228, 229, 230, 259, 270, 271, 272, 273, 274, 275, 277, 287
community members, 27, 35, 37, 46, 48, 66, 71, 78, 88, 89, 90, 101, 103, 106, 121, 165, 210, 225, 273
community needs, 71, 269
conference moderators, 99, 147, 165, 225, 259
Conference Rules, 216, 272, 274
cross-posting, 73, 78, 85, 201
cultural identity, 23, 78

D

deviance, 80, 167, 168, 169, 186, 192, 195, 209, 210, 225, 228, 229, 275
Deviant Behavior, v, 167, 286

E

Echo conference areas, 91, 147, 151, 259, 271, 282
email, 16, 17, 18, 33, 35, 59, 73, 82, 83, 84, 88, 101, 133, 141, 198, 200, 205, 209, 234, 243, 271, 283
emoticons, 4, 24, 26, 27, 28, 42, 160
ethnicity, 227, 228, 268
etiquette, 10, 24, 35, 147, 149, 167, 241, 246, 283

F

face-to-face interaction, 9, 22, 46, 47, 66, 269, 270, 271, 273
FAQ, 15, 16, 23, 24, 33, 36, 37, 42, 71, 72, 73, 74, 78, 81, 91, 121, 122, 123, 124,

125, 126, 194, 196, 198, 226, 242, 255, 277, 288, 292, 294
Fidonet, 11, 13, 16, 17, 18, 37, 38, 39, 40, 41, 42, 87, 88, 89, 90, 91, 95, 96, 99, 100, 101, 147, 154, 160, 164, 165, 209, 210, 211, 213, 214, 216, 218, 219, 225, 256, 257, 258, 259, 263, 267, 271, 272, 274, 281, 282, 283, 293
Fidonet History Project, 40
FidoNews, 39, 40, 42, 87, 91, 101, 147, 289, 292
finger, 70, 117, 118, 271
finger commands, 70, 117, 271
flame wars, 82, 125, 126, 137, 205, 209, 282
flaming, 37, 41, 126, 136, 140, 204, 205, 208, 215
flooding, 25, 52, 61, 173, 174, 195, 196, 198, 209, 214

G

gatekeeping, 80
gender, 117, 227, 228
generally stupid, 219, 220
group creation, 74, 75, 78, 243
group norms, 84, 103, 253

H

Handles, 93

I

identity information, 164, 271
interactions, 1, 4, 11, 47, 70, 73, 79, 91, 114, 116, 124, 134, 137, 140, 178, 179, 193, 216, 221, 236, 239, 255, 259, 263, 265, 270, 271, 272, 273, 274, 275, 282
International Coordinator, 88, 211, 256
Internet Protocol, 15, 277, 278
Internet Relay Chat, 13, 14, 23, 48, 105, 169, 230, 277, 281, 282, 283, 284, 292
Internet Relay Chat Community, 23, 48, 105, 169, 230

intimate bonds, 63
IRC network, 14, 15, 23
IRC protocol, 15

K

kill command, 232
kill file, 33, 198, 199, 200, 205, 209
killfile, 86, 152, 272

L

Lurker, 33, 282

M

mailbombs, 196, 198
male identity, 118
Matrix, v, 1, 2, 6, 7, 8, 9, 11, 13, 15, 19, 200, 269, 270, 273
Matrix Communities, v, 19
membership, 20, 27, 29, 45, 79, 80, 90, 147, 165, 231, 273
Mindless Chatter, 92, 95, 96, 99, 100, 215, 216, 217, 265
moderated newsgroup, 73, 75, 81
moderators, 81, 82, 86, 99, 101, 121, 147, 151, 165, 215, 221, 256, 259, 260, 263, 265, 267

N

net disruption, 50
Net.God, 33
Net.Legends FAQ, 33, 292
net-abuse, 36, 81, 193, 194, 195, 196, 197, 198, 226, 242, 292
netiquette, 10, 24, 28, 33, 34, 35, 36, 37, 41, 72, 73, 78, 123, 169, 172, 206, 242, 246, 247
netsex, 63, 115, 181, 182
Netsplit, 27, 50, 283
Network Coordinator, 88, 210, 211, 257, 258

newbies, 25, 26, 29, 58, 71, 72, 73, 91, 92, 99, 105, 106, 107, 109, 110, 114, 117, 120, 122, 124, 126, 130, 146, 147, 149, 150, 151, 152, 165, 170, 198, 216, 230, 233, 234, 235, 244, 254, 255, 256, 259, 260, 263, 267, 272, 283
newsgroup, 8, 10, 16, 31, 32, 33, 71, 72, 74, 75, 76, 77, 78, 79, 80, 81, 82, 83, 85, 86, 90, 91, 121, 122, 124, 125, 126, 128, 129, 130, 131, 132, 134, 136, 140, 141, 143, 146, 160, 165, 193, 195, 196, 198, 199, 200, 204, 205, 206, 209, 242, 245, 246, 247, 250, 253, 256, 282
newsgroup integrity, 77
nick, 25, 26, 66, 67, 69, 70, 106, 109, 114, 115, 116, 117, 118, 119, 120, 182, 183, 186, 188, 268, 279
NodeList, 88, 89, 90
nodes, 14, 17, 18, 38, 88, 89, 95, 256, 257, 258
norms, 11, 22, 24, 31, 37, 40, 43, 45, 47, 48, 52, 59, 61, 72, 78, 81, 86, 105, 120, 121, 125, 126, 127, 128, 130, 143, 150, 166, 167, 169, 170, 172, 192, 193, 194, 195, 198, 201, 202, 209, 213, 219, 225, 228, 229, 230, 233, 240, 247, 253, 255, 259, 267, 269, 271, 273, 274, 275, 283

O

off-topic posts, 91, 210, 213, 221
On-Line Communities, v, 227, 269
On-Line Self, v, 103
overloading nodes, 214

P

parent identity, 271
Personal Identity, v, 103
postings, 35, 36, 43, 72, 73, 78, 80, 81, 82, 86, 146, 194, 200, 250
posts, 33, 74, 79, 81, 83, 84, 92, 94, 101, 123, 124, 128, 132, 138, 143, 146, 160, 166, 193, 194, 195, 197, 198, 199, 200, 202, 203, 204, 205, 206, 208, 210, 213, 214, 216, 218, 219, 220, 221, 226, 243, 244, 247, 250, 254, 272, 282
private channel, 63, 65, 70, 105, 109, 110, 114, 120, 132, 166, 177, 178, 192, 198, 271
problem-solving strategies, 47, 48, 57, 91, 269
profanity, 24, 76, 97, 123, 172, 176, 186, 209, 213, 214, 240, 252
public channel, 63, 65, 106, 109, 112, 114, 115, 132, 166, 176, 181, 182, 205, 231, 240
punishment, 121, 127, 128, 129, 150, 167, 172, 195, 213, 257

R

race, 227, 228
real community, 1, 2, 3, 9, 269
reflexivity, 103
Regional Coordinator, 39, 88, 210, 211, 256, 257, 258
renegade, 81, 197, 242
Request For Discussion, 75, 82, 278
ridicule, 66, 179, 183, 192, 206, 209, 225

S

sarcasm, 26, 35, 192, 209, 218, 225, 279
sexual identity, 63, 117
signature files, 72, 73, 81, 83, 86, 99, 121, 143, 146, 271
sigs, 143, 198
social behavior, 48
social constraints, 242
social control, 34, 47, 71, 73, 80, 126, 143, 149, 168, 173, 178, 192, 197, 198, 199, 200, 201, 205, 209, 210, 211, 214, 218, 225, 230, 232, 272, 274, 275, 281, 283
social control mechanisms, 192, 201, 209, 225, 272
social distance, 232, 233, 234
social groupings, 19
social groups, 14, 47, 168
Social Inequality, v, 227

social institutions, 24, 47, 87, 269
social order, 35, 45, 46, 47, 49, 63, 66, 78, 82, 83, 87, 92, 103, 104, 105, 120, 143, 146, 152, 164, 165, 167, 168, 169, 177, 197, 200, 211, 221, 225, 236, 271, 272, 282
social rules, 88, 110, 167, 168, 169
Social stratification, 227
Socialization, v, 103
spam, 80, 101, 193, 194, 195, 196, 226, 242
spamming, 193, 194, 195, 196, 197
standardized message formats, 81, 86
stereotyping, 115, 259
stratification, 227, 228, 229, 230, 232, 233, 240, 241, 244, 256, 259, 267, 275
sysadmin, 196
sysops, 17, 34, 38, 39, 40, 42, 80, 81, 82, 86, 89, 90, 91, 121, 147, 165, 210, 212, 241, 242, 244, 256, 257

T

tagline, 99, 101, 220
Taglines, 99
TCP/IP, 15, 278
Technology, v, 13, 286, 291
Third places, 30
transgressions, 192
troll, 34, 36, 37, 86, 199, 221, 224, 250, 253
true name, 83, 89, 100, 118, 271

U

Undernet, 15, 23, 24, 27, 28, 29, 30, 40, 48, 49, 50, 51, 52, 55, 57, 63, 70, 71, 78, 82, 87, 90, 91, 99, 101, 105, 114, 116, 119, 120, 126, 130, 132, 147, 151, 170, 172, 176, 182, 192, 193, 195, 205, 209, 214, 219, 225, 230, 231, 232, 233, 235, 240, 241, 242, 244, 255, 256, 259, 267, 284
Undernet opers, 232, 242
Undernetters, 24, 42, 83, 109, 110, 114, 120, 130, 140, 141, 143, 147, 165, 192, 195, 198, 200, 205, 209, 213, 231, 241
Usenet, 8, 10, 11, 13, 15, 16, 17, 18, 31, 32, 33, 34, 35, 36, 37, 38, 40, 41, 42, 71, 72, 73, 74, 75, 78, 79, 80, 81, 82, 84, 85, 86, 87, 89, 90, 91, 99, 101, 121, 126, 128, 130, 132, 137, 139, 140, 141, 143, 144, 146, 147, 151, 154, 160, 164, 166, 193, 194, 195, 196, 197, 198, 199, 201, 202, 205, 209, 210, 214, 218, 225, 241, 242, 243, 244, 246, 247, 255, 256, 259, 267, 268, 271, 272, 274, 281, 282, 283, 284, 288, 290, 294, 295
User Group, 49
UUCP, 15, 16, 144

V

Virtual communities, 3, 78

W

work identity, 271

Z

Zone 2 War, 40
Zone Coordinator, 88, 211, 256, 258